MAPS OF MEANING

'Maps of meaning' refers to the way we make sense of the world, rendering our geographical experience intelligible, attaching value to the environment and investing the material world with symbolic significance. The book introduces notions of space and place, exploring culture's geographies as well as the geography of culture. It outlines the field of cultural politics, employing concepts of ideology, hegemony and resistance to show how dominant ideologies are contested through unequal relations of power. Culture emerges as a domain in which economic and political contradictions are negotiated and resolved.

After a critical review of the work of Carl Sauer and the 'Berkeley School' of cultural geography, the book considers the work of such cultural theorists as Raymond Williams, Clifford Geertz and Stuart Hall. It develops a materialist approach to the geographical study of culture, exemplified by studies of class and popular culture, gender and sexuality, race and racism, language and ideology. The book concludes by proposing a new agenda for cultural geography, including a discussion of current debates about post-modernism.

Maps of meaning will be of interest to a broad spectrum of social scientists, especially social and cultural geographers and students of cultural studies.

Peter Jackson is Professor of Human Geography at Sheffield University.

MAPS OF MEANING

An introduction to cultural geography

Peter Jackson

Routledge
Taylor & Francis Group
LONDON AND NEW YORK

First published 1989 by Unwin Hyman Ltd

Published 1992, 1994, 1995
by Routledge
2 Park Square, Milton Park, Abingdon, Oxon OX14 4RN
605 Third Avenue, New York, NY 10017

Routledge is an imprint of the Taylor & Francis Group, an informa business

Typeset in 10/12pt Bembo by Columns, Reading, Berks

British Library Cataloguing in Publication Data
A catalogue record for this book is available from the British Library

Library of Congress Cataloguing in Publication Data
A catalogue record for this book is available from the Library of Congress

ISBN 13: 978-0-415-09088-9 (pbk)

Contents

Foreword

The intellectual scene is changing fast. Concepts of place, space and landscape have become central to some of the most exciting developments across the whole field of the humanities and the social sciences. Where historians and anthropologists once studied individual actors and isolated communities, they now seek to place people in a shifting web of interdependencies which often stretches across the globe. Where economists and sociologists once constructed spaceless models of economies and societies, they now seek to account for the uneven development of capitalism and to make sense of the complex character of social life as it unfolds over space. Where political scientists once studied states as unitary actors or empty abstractions, they now seek to examine their territorial structures and to chart their changing involvements in inter-state systems. Philosophers and intellectual historians are alert as never before to the significance of 'local knowledge' and to the wider contexts in which their arguments move. And where human geographers once borrowed wholesale from other disciplines, they are now – as part and parcel of these changes – making major contributions in their own right.

Contours aims to introduce students to these extraordinary changes: in effect, to map the new intellectual landscape and help them locate their own studies within its shifting boundaries. We have tried to identify the most important issues, which are often the most interesting as well, to clarify what is at stake in the debates that surround them (without oversimplifying the arguments), and to illustrate what they mean in practical terms. In our experience most introductions leave the latest developments until last. These books mark a significant departure. They are written from the research frontier; they don't duck the difficult questions and neither do they reserve them for some future discussion. These are testing times for the humanities and the social sciences on both sides of the Atlantic, and we don't think there is anything to be gained by reticence of this kind. The ideas with which we are concerned in these books are of vital importance for *anyone* standing on the threshold of the twenty-first century. Living in multicultural societies in an interdependent world, in which events in one place are caught up in rapidly extending chains of events that span the globe; depending upon an increasingly fragile and volatile physical environment whose complex interactions require sophisticated analysis and sensitive management; recognising that the human impact on the face of the Earth has become ever more insistent – we have no choice but to enlarge the geographical imagination.

Contours is different in another way too. It is the product of a continuing series of discussions between all the authors involved. These are not books by committee, and we have all written what we wanted to write. But our arguments have been hammered out, revised and defended in regular meetings; every chapter in every book has been discussed by everyone; and ideas have constantly sparked across from one book to another. The result has not been consensus – nor was it supposed to be – but our views have often been changed by our discussions and we have all gained an increased respect for the opinions and approaches of others. Certainly, each of us has a different perspective on a different field, and we do not all think (or write!) in the same way. But we hope that each of these books conveys something of the excitement we have felt at working together: and the fun too.

Derek Gregory
Vancouver, April 1989

Preface

Described by Raymond Williams as one of the two or three most complicated words in the English language 'culture' defies easy definition. Anthropologists have used the term in literally dozens of ways, while geographers have proved no more adept at reaching a common understanding. At its most deceptively simple, 'culture' refers to the artistic and intellectual product of an élite. More generally, it refers to a system of shared beliefs or a whole way of life. Rather than being a source of confusion, however, the very fact that 'culture' is a contested term is a vital key to its understanding. For 'culture' is not the safe preserve of an élite who dominate a country's major cultural institutions and define its 'national culture'. It is a domain, no less than the political and the economic, in which social relations of dominance and subordination are negotiated and resisted, where meanings are not just imposed, but contested.

This book employs a more expansive definition of culture than that commonly adopted in cultural geography. It looks at the cultures of socially marginal groups as well as at the dominant, national culture of the élite. It is interested in popular culture as well as in vernacular or folk styles; in the urban as well as the rural; in contemporary landscapes as well as relict features of the past. Drawing on the literature of cultural studies and social theory, it introduces a variety of new perspectives on the geographical study of culture besides the landscape approach to which cultural geographers (particularly in North America) remain so devoted, still under the thrall of Carl Sauer and his influential 'Berkeley School' (see Ch. 1).

The achievements of the 'Berkeley School' notwithstanding, its continuing influence throughout human geography conceals a number of deficiencies inherent in Sauer's approach to culture. Chapter 2 considers these problems and proposes a materialist alternative, taking its inspiration from Raymond Williams' studies in cultural criticism. Williams' work sought to clarify the relations between culture and society by employing the Marxist categories of ideology and hegemony (reviewed in Ch. 3). The next four chapters exemplify the application of a cultural studies approach to some contemporary issues in human geography concerning class and popular culture (Ch. 4), gender and sexuality (Ch. 5), 'race' and racism (Ch. 6), and language (Ch. 7). The book concludes by sketching an agenda for cultural geography (Ch. 8), based on the notion of cultural politics advanced in the introduction.

Rather than trying to be comprehensive in its approach to cultural geography, the book provides a preliminary survey of some major themes. To take just two of the most obvious omissions; there is little discussion of advertising or the media, despite their pervasive influence and the existence of innovative geographical work in this field (Burgess & Gold 1985); and there is only passing reference to the geography of religion despite recent attempts to clarify the theoretical foundations of this particular branch of cultural geography (Levine 1986). With the rise of Islamic fundamentalism, the growth of the 'electronic church' in the United States, and continuing sectarian struggles in Northern Ireland, the contemporary significance of religion and its geographical expression can scarcely be exaggerated. But with relatively few exceptions, including some important studies of 'territoriality' and residential segregation in Belfast (Boal 1969, Boal *et al.* 1977), there has been little innovative work by geographers in this field.

The book is also heavily weighted towards British and American examples where most of my own research has been based. These are the societies with which I am most familiar and one has to start somewhere. But the absence of other parts of an increasingly interdependent world should not pass without comment. Indeed, at one level, this book is an attempt to deal with my own place in the contemporary world, writing as a white, middle-class man, working in a privileged though increasingly beleaguered profession and living in the capital of what was once the heart of Empire. I hope, at least, that readers will not infer that British and American examples predominate because they are assumed to be inherently superior or more important than any others. If social science teaches us anything, it should be an enthusiasm for cultural difference and a willingness to see that our own society represents just one way of doing things among a wide range of possibilities. Every study in cultural geography is, at least by implication, a comparative one and we should, wherever possible, be explicit about the problems that such relativism entails.

By taking a few themes, this book attempts to illustrate the potential for a revitalized cultural geography, drawing on theoretical developments in cultural studies and social theory and informing that work with a more sophisticated geographical sensibility. To some extent, then, this book is an extension of my earlier work with Susan Smith which sought to explore some of the theoretical foundations of contemporary social geography (Jackson & Smith 1984). But it is novel in its attempt to retheorize culture and to suggest new ways of approaching that concept geographically. I will argue that geography is not merely incidental to cultural variation, relevant only to the explanation of diversity, but that it is fundamental to the very constitution of culture. If social processes do not take place on the head of a pin, then we need to take spatial structure seriously, not least in the production and communication of meaning that we call culture.

Although in one sense this book is a solo effort, it is also the product of collaboration with a team of highly supportive colleagues. Publication of this volume inaugurates a series that was initiated by Derek Gregory, goaded into action by Mark Cohen, and brought to fruition by Roger Jones. Other members of the team also deserve thanks for providing a congenial yet critical writing environment: Morag Bell, Felix Driver, Roger Lee, David Livingstone, Graham Smith, Nigel Thrift and Peter Williams. Paul Richards provided inspiration at the start of this project while Denis Cosgrove dissuaded me from overhasty publication. I also benefited from designing and then teaching a new course on cultural geography at UCL with Jacquie Burgess and Hugh Prince. In the course of writing this book I enjoyed a particularly productive winter quarter in the United States in 1987 on a teaching exchange with the University of Minnesota and Macalester College in Minneapolis–Saint Paul. Thanks especially to David Lanegran (who instituted the exchange) and to Helga Leitner, Eric Sheppard, and members of the reading group that met in Eric and Helga's house. Many people have been generous enough to read and comment on earlier drafts of this book or to respond to seminar and lecture presentations. They include Bob Catterall, Hugh Clout, Andrew Crowhurst, Tim Cresswell, Richard Dennis, Peter Goheen, Jane Jacobs, David Ley, Jo Little, Miles Ogborn, Lisa Popik, and David Ward. Special thanks to Sarah Whatmore who read (and re-read) sections of the manuscript, often under pressure. Conferences at UCL, UBC and Phoenix, Arizona all helped clarify my ideas as did visits, near the end of the project, to Adelaide and Sydney. The maps were expertly drawn (or redrawn) by Lauren McClue, and Nina Laurie gave valuable editorial assistance. Thanks to all concerned, including those students who have heard all this before. Despite their assistance, any remaining defects are my own responsibility.

Peter Jackson
University College London

Acknowledgements

We are grateful to the following individuals and organisations for permission to reproduce photographic illustrations (figure numbers in parentheses):

Goldsmiths' Library, University of London (4.5); Harvard Theatre Collection (6.3); Kunsthistorisches Museum, Vienna (4.1); Jacob A. Riis Collection, Museum of the City of New York (4.4); Richard Symanski (5.1, 5.2, 5.3,); Tate Gallery Publications (4.7); the Trustees of the British Museum (6.1); University of California Press (1.1); and the Welsh Arts Council (2.1).

It has unfortunately not proved possible to trace the copyright for Figures 3.2, 4.6, 6.2, 6.4, 6.5, and 6.6.

The following illustrations have been redrawn in the Cartographic Unit at UCL from the sources noted: Figs. 2.2 & 2.3 (Pugin, 1836); 3.4 (Ley, 1974); 3.5 (Ley & Cybriwsky, 1974); 3.6 (Cohen, 1972); 4.2 & 4.3 (Bell & Bell, 1972); 5.4 (Levine, 1979); 5.5 (Castells, 1983); and 7.2 (Stallybrass & White, 1986).

All remaining illustrations are original.

The following publishers and authors have kindly granted permission for the reproduction of text extracts: Alfred A. Knopf, Inc., for permission to reproduce extracts from *City of women* (C. Stansell); International Thomson Publishing Services Ltd., for permission to reproduce extracts from *Orientalism* (E. Said); Oxford University Press (New York), for permission to reproduce an extract from *Blacking up* (R. Toll).

MAPS OF MEANING

Introduction: maps of meaning

The current transformation of cultural geography is taking place as a result of its dialogue with social geography and cultural theory. Until recently, its range was limited to the interpretation of historical, rural, and relict landscapes, and to a static mapping of the distribution of culture traits, from barns and cabins to field systems and graveyards. Emerging from its antiquarian phase, cultural geography has begun to assume a more central position in the current rethinking of human geography. Cultural geographers are now experimenting with a range of new ideas and approaches, their aversion to theory now firmly overcome. These developments have drawn extensively on contemporary cultural studies and on other theoretical developments across the social sciences. But the traffic has not all been in one direction: there is now at least the potential for repaying this debt by informing cultural studies with some of the insights of social and cultural geography. This book surveys some of the most significant recent developments in cultural theory and outlines the agenda for a theoretically informed cultural geography. This reorientation involves a growing convergence of interests between those with an historical interest in the evolution of geographical landscapes and those with a contemporary interest in cultural studies and social theory.

Though much of the 'new' cultural geography remains wedded to the idea of landscape, the approach adopted here emphasizes the *plurality of cultures* and the multiplicity of landscapes with which those cultures are associated. It rejects a unitary view of culture as the artistic and intellectual product of an élite, asserting the value of popular culture both in its own terms and as an implicit challenge to dominant values. Culture emerges as a domain in which economic and political contradictions are contested and resolved. This does not mean that the cultural is reduced to its political and economic determinants. Neither does it mean that culture can be dismissed as ephemeral: a residual category left unexplained by more rigorous analyses of political economy. Rather than analyzing each of these domains in isolation, it puts the relationship between culture and society at centre stage, insisting on the relative autonomy of the cultural and exploring its specific intersections with the political and economic. This involves a shift in emphasis from culture itself to the domain of *cultural politics* where meanings are negotiated and relations of dominance and subordination are defined and contested. Rather than separating 'culture' from 'politics', therefore, this book is concerned with the extent to which *the cultural is political*. As many of these ideas have emerged from cultural studies, a brief survey of developments in that interdisciplinary field provides a useful starting point.

An introduction to cultural studies

The postwar revival of cultural studies in Britain began with the work of two eminent cultural critics: Richard Hoggart, the author of *The uses of literacy* (1957) and a tireless champion of popular (working-class) culture; and Raymond Williams, a Marxist Professor of Drama at Cambridge who, in his many works, consistently emphasized the connections between culture and society (Williams 1958). Their work was developed by members of the Centre for Contemporary Cultural Studies at the University of Birmingham, under the directorship of Stuart Hall, now at the Open University.[1] The Centre undertook work in a variety of fields, from language and media studies to the analysis of race and gender. Characteristic of their work was an emphasis on ethnography (detailed qualitative fieldwork) as well as on the development of theory, and an intellectual practice that emphasized collective work over individual scholarship.

Hall's colleagues at the Centre for Contemporary Cultural Studies have provided a working definition of culture as the level at which social groups develop distinct patterns of life. Culture refers to 'the way, the forms, in which groups "handle" the raw material of their social and material existence' (Clarke *et al.* 1976, p.10). Elaborating slightly on this, 'culture' refers to the codes with which meaning is constructed, conveyed, and understood. Significantly, authors at the Centre employed a geographical metaphor to describe this process, and one that is singularly appropriate to the theme of this book: cultures are *maps of meaning* through which the world is made intelligible.[2] Cultures are not simply systems of meaning and value carried around in the head. They are made concrete through patterns of social organization. Culture is 'the way the social relations of a group are structured and shaped: but it is also the way those shapes are experienced, understood and interpreted' (ibid., p.11).

Cultures therefore also involve relations of power, reflected in patterns of dominance and subordination. Dominant cultural institutions, such as the BBC or the *New York Times*, Henley Regatta or Royal Ascot, exert a subtle and pervasive influence on the lives of many thousands of people, establishing a 'preferred reading' of local and national circumstances. This is not to imply that the state imposes 'social control' in a direct or sinister manner. Rather, it suggests that dominant views are most effective if they become 'naturalized' as part of everyday common sense. The concepts of ideology and hegemony are therefore central to ıcultural studies, referring to the proceses through which dominant meanings are imposed, negotiated, and registered.

Hall and his colleagues continue to suggest a way of dealing with the plurality of cultures that exist in any complex society such as contemporary Britain. Cultures, they argue, are ranked hierarchically in relations of dominance and subordination along a scale of 'cultural power' (Clarke *et al.* 1976, p.11). Subordinate cultures frequently appropriate material resources from one domain and transform them symbolically into another: the Crombie overcoat and razor-cut of the Skinhead, for example, exaggerate and transform the symbols of working-class respectability as stylized items of protest and insubordination. 'Rituals of resistance' of this kind are a staple of contemporary cultural studies (Hall & Jefferson 1976, Hebdige 1979). Tracing the material circumstances in which such transformations occur (see Ch. 3) is one of the central tasks of a theoretically reconstituted cultural geography.

This book combines some of the most important ideas from cultural studies with some recent developments in human geography, seeking alternative approaches to the geographical study of culture from the traditional obsession with landscape. Alternatives include the 'society and space' debate (with its eclectic mix of structuration theory, realist philosophies of science, Marxism, and time-geography); the concept of spatial divisions of labour; and the theory of uneven development (see Ch. 8). These ideas have already made a considerable impact in economic and social geography while cultural geographers, with their steadfast aversion to social theory, have remained relatively immune from them.

Culture is too important a domain to be left to the specialist in cultural geography. If geographers are to make the most of recent developments in social theory, for example, they require a more sophisticated theory of culture. For culture is not only socially constructed and *geographically expressed*. Following the much heralded reassertion of space in critical social theory (Gregory & Urry 1985, Soja 1989), it must also be admitted that culture is *spatially constituted*. To take some examples from the following chapters, it will be shown that gender relations are differently constituted in different labour markets (see Ch. 5); that racism takes a different form in different localities and at different times, changing shape according to changing historical and geographical circumstances (Ch. 6); that the emergence of San Francisco as a Mecca for gay men and the development of gay politics in that city has a clear spatial basis and a precise territorial form (Ch. 5); and that the shifting boundaries between public and private space in 19th-century London and New York gave rise to serious problems of 'social control' (Ch. 4).

These examples are all concerned with the politics of culture. The next section develops this theme and speculates on the importance of the political context in explaining the current resurgence of interest in cultural studies among geographers and other social scientists.

Cultural politics and the politics of culture

That the cultural is political follows logically from a rejection of the traditional notion of a unitary view of 'culture', and from a recognition of the plurality of cultures. If cultures are addressed in the plural (high and low, black and white, masculine and feminine, gay and straight, urban and rural) then it is clear that meanings will be contested according to the interests of those involved. Consider the following attempts to define a national culture, the first from T.S. Eliot, the second from Hanif Kureishi. For Eliot, British culture 'includes all the characteristic activities of a people: Derby Day, Henley Regatta, Cowes, the twelfth of August, a cup final, the dog races, the pin-table, the dartboard, Wensleydale cheese, boiled cabbage cut into sections, beetroot in vinegar, nineteenth-century Gothic churches and the music of Elgar' (Eliot 1948, p.31). For Kureishi, the list would be radically different and would include: 'yoga exercises, going to Indian restaurants, the music of Bob Marley, the novels of Salman Rushdie, Zen Buddhism, the Hare Krishna Temple, as well as the films of Sylvester Stallone, therapy, hamburgers, visits to gay bars, the dole office and the taking of drugs' (Kureishi 1986, pp.168-9). Far from confirming Eliot's fears about the homogenizing tendencies of 'mass culture', Kureishi's list suggests that local, urban, and regional cultures are more distinct than ever, reflecting their changing social geography. The tension between 'high' and 'low' cultures identified in Eliot and Kureishi also exists between those who fear that regional cultures are being eliminated by the globalizing tendencies of capital accumulation (Peet 1986) and those who champion the active capacity of subordinate cultures to resist and subvert those tendencies, celebrating the persistence of 'cultures of difference' (Clarke 1984). The 'geography of culture' is itself a contested terrain.

In advocating a cultural politics, one cannot ignore the politics of culture. The present reformulation of cultural geography should therefore be situated in its immediate political context. What is there about the current politics of fiscal retrenchment, privatization, and economic recession in Thatcher's Britain or Bush's America that might be relevant to a revival of interest in cultural studies? Why have such phrases as 'enterprise culture', 'Victorian values', and 'moral majority' gained such sudden salience? Is the age of the yuppie and corporate culture, of urban heritage and rural nostalgia, of football hooliganism and inner-city rioting, a response to national economic decline (as Wiener 1981, Walvin, 1986, and Hewison 1987 each maintain)? Or does it not also represent the growing confidence of the 'consumption classes' and the increasing alienation of the impoverished and despairing 'underclass', each with its own distinctive geography? An understanding of contemporary national culture clearly necessitates an appreciation of these changing political and economic contours.

The contours of contemporary culture include a paradoxical mixture of trends towards the general and the particular. On the one hand is the widespread emergence in the high street of international fast food stores, like McDonald's and Pizza Express, and their equivalents in clothing and other commodities. On the other hand is the attempt to match every new product to increasingly specific market 'niches'. The fields of 'commodity aesthetics' (Haug 1986) and 'life-style advertising' (Mort 1988) have emerged to provide ways of differentiating consumers from one another, maximizing the appeal of each new product by associating it with an appropriate life-style while simultaneously emphasizing the consumer's individuality and personal taste. A similar paradox applies to the new technology with its tendency simultaneously to liberate and enslave. Some people decry the insidious way in which the new technology colonizes new domains, with cable TV and satellite dishes bringing an endless succession of standardized images into the home. The personal stereo, the video, and the word processor have likewise blurred the boundaries between work and leisure, public and private space. But others argue that new technology provides users with the potential to get 'inside the machine' (Chambers 1986) to produce new audio-visual combinations, personalized programming, a fresh montage of sounds and images.

Contemporary music provides an example of both tendencies. Its populist, democratizing potential is present in the endless reworkings of dub and talk-over, cutting and mixing, scratch, rap and hip-hop (Hebdige 1987). But it can also be seen as a threat to more 'authentic' forms of musical production – anyone with a turntable, a cassette recorder or a DX–7 synthesizer can be an instant musician. The new technology also threatens to overwhelm its audience with a cacophony of sounds and images, as frustrated viewers endlessly switch TV channels by remote control or via the fast-forward and reverse buttons on their VCRs. It is not so much that technology dictates a particular pattern of social relations, but that its reception is socially and culturally mediated. As Chambers argues, it is part of the post-modern condition to find ourselves 'walking a narrow line between the enlargement of meaning and the peril of it breaking down and evaporating altogether' (1986, p. 199). But it is also a familiar cultural reaction to the advent of new technology as 19th-century reactions to the 'mechanical poison' of photography attest (ibid. p.72).

As these brief examples show, the politics of consumption are rarely straightforward. For white, middle-class teenagers, listening to the Bhundu Boys or to the latest Bhangra band can seem more like 'musical tourism' than a genuine expansion of musical consciousness. For the history of black music is a history of exploitation and appropriation. The cultural politics of 'Band-Aid' or 'Graceland' are even more complex: did they raise people's consciousness in the West and increase Third World emancipation or were they an extremely sophisticated form of self-indulgence, voyeurism, and exploitation? These political tensions within contemporary culture are part of the current debate about post-modernism and the confusion of meanings that attend it. Every message is capable of multiple readings. Every account bears the impress of multiple authors; and no single intention can be inferred. These political dilemmas lift post-modernism above the epiphenomenal and demand that it is properly conceptualized as a social process, periodized in terms of production as well as consumption (Zukin 1988a).

As an extension of this argument about the politics of contemporary culture, it is no coincidence that the current revitalization of cultural studies is taking place at a time when various aspects of Britain's cultural diversity are under threat from an increasingly intolerant and authoritarian government. The last couple of years have seen gay rights under attack from proposed changes in local government legislation; women's rights to legal abortion have been threatened with severe curtailment; anti-racist initiatives have been jeopardized by events in Bradford and Manchester; and educational freedom has been completely redefined by the abolition of tenure, accusations of political bias in schools, and the imposition of a common national curriculum. Similar trends can be discerned in the United States with the rise of the 'moral majority'; the renewed virulence of racism (at Howard Beach in New York, for example, and on university campuses throughout the country); the devastating effects of the AIDS crisis on the gay community; and the repeal of the Equal Rights Amendment in many states. Significantly, too, in such a depressing political climate, many of the most optimistic developments have occurred in the field of cultural politics. The progressive agenda of Jesse Jackson's 'Rainbow Coalition', for example, has sought to define common ground between blacks and Hispanics, lesbians and gay men, urban and rural poor, and other oppressed groups.

For some Marxist writers, however, the current revival of interest in cultural studies is interpreted as a diversion of intellectual energy away from the more pressing questions of political-economy. Formerly radical authors stand accused of abandoning their commitment to radical social and economic change, substituting a softer cultural analysis for the harsher realities of class struggle. Such an interpretation can be applied to David Harvey's recent attack on urban studies, accusing its practitioners of 'a marked strategic withdrawal from Marxist theory', 'an abrogation of scientific responsiblity' and 'a caving in of political will' (Harvey 1987a, pp.367, 376). Harvey's refusal to abandon the 'tough rigour' of dialectical theorizing is paralleled by Neil Smith's hostile critique of recent locality studies, with their sensitivity to 'regional cultures', in which he detects a dangerous 'empirical turn' (Smith 1987a). These interventions have led to a reconsideration of the politics of social and cultural theory, a theme which must be central to any redefinition of the field of cultural geography.

But there is nothing inherently conservative about cultural studies, even among those who choose to examine 'élite' sources, as Raymond Williams' work proudly attests (Eagleton 1988). While cultural studies may be dismissed by some people as a reactionary diversion, to others it offers an important domain for political debate, having provided new grounds for collective struggle. Cora Kaplan's work provides a model here, suggesting that struggles around cultural definitions of gender and race have generated much political energy during the 1980s (Kaplan 1986, p.6). But neither Greenham Common nor the Brixton riots have eclipsed traditional forms of class struggle. Traditional struggles have simply been expressed in other ways and with unpredicted consequences. For example, a 'typical' working-class confrontation, such as the 1984–5 miners' strike, had significant 'cultural' effects, not least in challenging the persistence of patriarchal gender relations.

One area in which culture may provide a haven for political reaction is in the scope it affords for imprecision and circumlocution. Debates about racism and anti-racism, for example, can be defused if they are represented as debates about 'multi-culturalism' where attitudes are less polarized and where the liberal demand for tolerance and fair play obscures deeper questions of inequality and racism. The current preoccupation within cultural studies with the analysis of language, discourse, and text, as opposed to the analysis of social action, lends itself to similar abuse. Sensitivity to language can be a good guide to political commitment, however. There is a world of difference, for example, between those who employ the liberal vocabulary of agency, context, and interaction, and those who prefer the more radical language of structure, power, and struggle. A sensitivity to the politics of language (see Ch. 7) is therefore a central component to any reworking of cultural studies in human geography.

To summarize, this book attempts to reformulate a theory of culture around the current *reapprochement* between social and cultural geography. It suggests that cultural geography must be contemporary as well as historical; theoretically informed yet grounded in empirical work: sympathetic to other conceptions of human geography rather than focused exclusively on landscape; and concerned with a range of cultures and with the cultural politics that this implies. Cultural geography can no longer be dismissed as 'a celebration of the parochial' or 'a contemplation of the bizarre' (Gregory & Ley 1988, p.116). As a serious intervention in the culture of modernity, the 'new' cultural geography has an insistently critical, political edge. This book is a contribution to that critique.

Notes

1 The development of the Centre, from the ideas of Hoggart and Williams, through Thompson, Gramsci, and linguistic structuralism, to the 'impact of the feminisms' and the need for more 'concrete studies' is described in a stimulating essay by Hall (1980a). A similar genealogy is traced by Chambers (1986, Ch.11).

2 Chambers (1986) also uses a variety of geographical metaphors, intending to provide 'a map of popular culture', a 'horizontal reading associated with maps' and a guide to 'the geography of the imagination' (a phrase also used by Davenport 1984). Hebdige (1988) goes even further in this direction, charting a 'cartography of taste', 'mapping out' sub-cultural styles and 'imagined territories', detailing 'stylistic terrains' and 'invasions of symbolic space'.

Chapter one
The heritage of cultural geography

Cultural geography is in urgent need of reappraisal; its conception of culture is badly outdated and its interest in the physical expression of culture in the landscape is unnecessarily limited. In trying to find a way round these problems, this book argues for a more expansive view of culture including its less tangible aspects such as those embodied in symbolic forms and in everyday social practice, and it explores a range of geographies besides those that focus exclusively on landscape. Before proceeding to introduce these new approaches, however, this introductory chapter reviews the current stasis of cultural geography by providing a critical survey of its origins and development, particularly in North America, where it is shown to have been the product of one particular school (the 'Berkeley School') and one remarkable man, Carl Sauer.

The chapter discusses the intellectual context of Sauer's work, including his liberal borrowing of concepts and ideas from the anthropologists Alfred Kroeber and Robert Lowie. His espousal of a 'super-organic' approach to culture is criticized and his inordinate influence on later generations of cultural geographers is traced. Finally some alternative conceptions of culture are introduced, to be discussed at length in subsequent chapters. Adoption of these alternatives involves a complete rethinking of cultural geography in which a convergence with social geography can be anticipated (Jackson 1980). The chapter begins, though, by demonstrating how the current stasis of cultural geography has arisen from its attachment to an outmoded conception of culture, inherited from the work of Carl Sauer and his colleagues at the Berkeley School.

The Berkeley School and its legacy

In the United States, 'cultural geography' is virtually a synonym for 'human geography', including several aspects of the subject that would be thought more appropriate to economic or social geography as they are currently practised in Britain. Introductory courses in cultural geography are taught to large classes of students, including many who have no intention of becoming geography majors. For many such students, it is their only contact with academic geography, the more so as geography is not commonly taught as a separate subject in high school. Several undergraduate textbooks have been designed for this market (e.g. Spencer & Thomas 1973, Jordan & Rowntree 1982, de Blij 1982). Insofar as they approximate the coverage of human geography in the British sense, their scope is correspondingly large. But in their definition of culture and its expression in the landscape, they seem excessively restrictive, as much in what they leave out as in what they include (cf. Norton 1984). Their content and approach owes much to the influence of Carl Sauer who, with the possible exception of Vidal de la Blache in France (Cosgrove 1983), occupies a unique place in the history of cultural geography.[1]

Carl Sauer (1889–1975) dominated North American cultural geography throughout his lifetime and particularly during his years as head of the influential Berkeley School, a position which he assumed at the age of 33 and which he held until three years before his retirement in 1957. At that time, as one of Sauer's students has remarked, geography at Berkeley was still less a department than an individual (Parsons 1979, p.9). During his time at Berkeley, Sauer supervised some 40 PhD theses, the majority on Latin American and Caribbean topics, conveying to all his students his firm belief in the need for first-hand field experience and for learning the language of the people being studied. A monolingual PhD was for Sauer a contradiction in terms (Sauer 1956a). Many of Sauer's graduate students went on to hold senior academic posts in their own right. Through them, Sauer continued to influence a second generation of American geographers.

Sauer was twice president of the Association of American Geographers (in 1941, and again in 1956), a position that gave him the opportunity to make a number of influential statements on the nature of the discipline.[2] In 1941, Sauer spoke on the nature of historical geography, protesting against its general neglect by his American colleagues. Geography in the United States was a native, Midwestern product, he argued, and its development was a faithful reflection of this fact, dispensing with any serious consideration of cultural and historical processes. Sauer argued for a broad definition of the subject of geographical inquiry, fiercely opposing all forms of academic pedantry, and deploring those who valued logic above intellectual curiosity. His style was a characteristic mixture of the avuncular and the iconoclastic:

> Only if we reach that day when we shall gather to sit far into the night, comparing our findings and discussing all their meanings shall we have recovered from the pernicious anemia of the 'but-is-this-geography' state (1941 p.4).

Despite this apparent catholicity concerning the subject matter of geography, Sauer nonetheless restricted his comments and the great majority of his own research effort to the material aspects of culture as expressed in the 'cultural landscape' (see below). It was this excessive focus on the material elements of culture and their representations in landscape that had such a profound influence on the development of American geography.[3]

In his second presidential address in 1956, Sauer elaborated on this educational philosophy. On this occasion, he spoke on the unspecialized quality of geography. The ideal undergraduate curriculum, he argued, would have a limited number of geography courses, enriched by courses in the liberal arts and especially in natural and cultural history. The geographer's best training came in the form of an active apprenticeship, he argued, doing fieldwork and developing the skills of experienced observation. He described the ideal field course as a peripatetic form of Socratic dialogue, a running exchange of questions between student and teacher, prompted by the changing scene. The most memorable portrait of Sauer shows him in such a characteristic pose, the accompanying text amplifying his mood (Fig. 1.1).

Despite his pre-eminence within American geography, Sauer felt himself to be rather out of step with his times. He withdrew from academic geography at a relatively early stage, championing the role of the individual scholar and opposing what he saw as the bureaucratization of social science research. There is a certain irony, then, in the extent to which Sauer's ideas reflect his own socialization within a particular academic milieu. Sauer's parents were of German ancestry and he himself spent several years at school in southern Germany. He was heavily influenced by the German cultural and historical sciences (*Geisteswissenschaften*), and, although he acknowledged the work of such British geographers as Vaughan Cornish and H. J. Fleure, and Americans like George Perkins Marsh, it was from the German classics (Ritter, Humboldt, Ratzel, and Hahn) that Sauer derived his perspective on culture and landscape.

After taking some graduate work in geology at Northwestern University, Sauer transferred to geography at Chicago. His intellectual debts to the geologists Rollin D. Salisbury and Thomas C. Chamberlin,

From photograph by K. J. Pelzer, September, 1935.

"Locomotion should be slow, the slower the better; and should be often interrupted by leisurely halts to sit on vantage points and stop at question marks."

Figure 1.1 Carl Sauer in pensive mood

whom he encountered at Chicago, have often been remarked (Parsons 1976, Entrikin 1984). From them, Sauer derived his model of scientific method, his evolutionary perspective, his commitment to an extended time-span (what he called 'the interesting far reaches of geologic time'), and his belief in the virtues of field research. His reading of the German geographers Ratzel, Schluter, and Hahn encouraged him to reject environmental determinism and to search for an alternative perspective on the human impact on the landscape. His geological sources provided little help in this search. Instead, Sauer turned to the German Romantics, and particularly to Goethe, whom Sauer admired for his rejection of the increasing specialization of modern science and its disregard of subjectivity and symbolism (Bowen 1981, Speth 1981). For, as Speth has argued, Goethe's conception of morphological change, with its dual emphasis on form and process, proved highly influential in the development of Sauer's own ideas on the cultural landscape.

The cultural landscape

Sauer advanced his most influential concept in a methodological paper called 'The morphology of landscape' (1925). It was this paper that, in the opinion of one of his students, 'catapulted Sauer into international attention' (Parsons 1979, p.13), although the same author recalls that Sauer was himself rather dismissive about the paper's reception, suggesting that several people seemed to have spent more time reading it than he had writing it. The paper began by defining 'landscape' as 'the unit concept of geography', a 'peculiarly geographic association of facts' (Sauer 1925, p.25). He then used this concept of landscape to describe 'a strictly geographic way of thinking of culture', that is, the impress of the works of man (*sic*) upon an area (ibid. p.30). The approach was later exemplified in the Foreword to *The early Spanish main* (1966) where Sauer drew attention to the geographical significance of the United States' southern boundary with Mexico. The international boundary, he argued, ran against the grain of the continent, from the Rio Grande to the Pacific coast. It was a cultural rather than a physical divide:

> The same mountains and deserts, pine forests, oak woodlands, scrub, and grasslands extend north and south; the difference is the people and their ways. On this side, change has been accelerating and innovation has become the dominant order of living. On the other side, ways of past experience and acceptance have been retained in gradual modifications (Sauer 1966. p.v).

The same physical environment has given rise to quite different cultural landscapes because of different cultural processes in each area. The cultural landscape was thus contrasted with the physical landscape, the former, in Sauer's classic phrase, having been 'fashioned out of a natural landscape by a culture group' (Sauer 1925, p.46).

Although the definition is not particularly contentious, except insofar as one might have difficulty specifying precisely what is meant by a 'culture group', little else in the paper can stand without comment. Sauer went on to argue, for example, that 'Culture is the agent, the natural area is the medium, the cultural landscape the result' (ibid. p.46). This attribution of agency to culture is highly problematic and is symptomatic of Sauer's teleological approach. By attributing causality to 'culture' rather than to particular individuals or social groups, Sauer implicitly diverted attention away from the social and towards the physical environment. Sauer's educational philosophy is relevant again here. He was resolutely opposed to the 'scholasticism' of social theory and equally strongly of the opinion that geographers should retain close contact with their colleagues in the natural sciences. This view extended even to his opinions on anthropology. Although he spoke warmly of the prospects for a gradual coalescence of social anthropology and geography as the first of a series of fusions into a larger science of man (ibid. p.53), he later came to adopt a more critical stance towards anthropology because of what he perceived to be its growing interest in social theory and social welfare (Entrikin 1984). In order to understand this apparent transformation in Sauer's ideas, it is worth reflecting on his conception of human agency, as revealed through his approach to 'culture history', before returning to Sauer's particular interpretation of cultural anthropology.

Culture history and human agency

Although Sauer was influential in setting up the international symposium on *Man's role in changing the face of the Earth*, entitling his own paper 'The Agency of Man on Earth' (Sauer 1956b), his principal interest was in landscape as a record of human activity rather than in the social systems through which human agency is actively expressed. Sauer defined agency as 'the capacity of man (*sic*) to alter his natural environment' (ibid. p.49). Although he spoke of agency in terms of 'historically cumulative effects', his discussion concentrated on physical and biological processes set in motion by human intervention rather than on social processes *per se*. Sauer provides a perspective on the environment as 'deformed', 'deflected', and 'appropriated' by human beings, with an implicit moral stand against the 'destructive exploitation' of the Earth's resources. It is not an argument about human agency in the contemporary sense of a capacity for progressive social change (Gregory 1981, S. W. Williams 1983).

There is, however, evidence of considerable equivocation in Sauer's thinking about the significance of human agency. From an early emphasis on morphology and a concomitantly passive conception of the scope for human agency, Sauer shifted his primary research focus to a more active appreciation of the social transformation of landscape. This change of emphasis is most clearly seen in Sauer's essay on 'Historical geography and the western frontier' (1929) and in his attempts to repudiate an earlier generation's belief in the determining effect of environmental influences. Much of his methodological writing was directed towards this end. But he tended to substitute for environmentalism an understanding of culture that was scarcely less constraining.

The model Sauer adopted for cultural geography was that of geology and the earth sciences rather than history and the humanities. He found the time-scale of the geologist particularly appealing and took an evolutionary approach to history. Although Sauer thought his approach to landscape essentially similar to that of the cultural historian, it was a very particular view of history that he espoused (cf. M. Williams 1983). For Sauer, the most intellectually engaging problem was the search for the origins of an institution or culture trait rather than an interest in the dynamics of social change. Coupled with his evolutionary perspective, Sauer also tended to see history unproblematically in terms of tradition. He viewed with regret the homogenizing tendencies of the modern world. Urban industrial society was decidedly not to his taste.

> By chance and choice I have turned away from commercialized areas and dominant civilizations to conservative and primitive areas. I have found pleasure in 'backward' lands, where the demands of industry for materials and markets are little felt (Sauer 1952, p.4).

Throughout his work, Sauer betrayed an anti-modernist tendency that went hand-in-hand with a fundamentally conservative outlook. Culture was equated with custom; cultural diversity as an unqualified good. Sauer's inherent conservatism never seems to have troubled his students for whom the 'Old Man's' ideas took on the status of common sense (Entrikin 1984, p.407).[4] The current agenda of cultural geography in the United States is still dominated by Sauer's original concerns with rural, vernacular and folk themes. While it shows a respect for tradition and a fascination with diversity, it also betrays a reactionary attitude towards social and cultural change, not least in terms of the agenda that is not addressed. The approach to culture from which this attitude derives can be traced back to Sauer's interest in cultural anthropology.

Cultural anthropology: Boas, Lowie, and Kroeber

Sauer once wrote of anthropology as methodologically 'the most advanced of the social sciences' (Sauer 1941, p.6). His respect was directed towards the physical rather than the social aspects of anthropology, and more specifically towards the uniquely North American tradition of cultural anthropology. This tradition began with the work of Franz Boas who trained initially as a geographer and shared Sauer's own commitment to first-hand field research (Trindell 1969). But it was two students of Boas, Alfred Kroeber and Robert Lowie, who had most influence on him. Lowie introduced Sauer to the second volume of Ratzel's *Anthropogeographie*, the book that signalled Sauer's final liberation from the conceptual straitjacket of environmental determinism. Lowie also drew Sauer's attention to Eduard Hahn's work on the domestication of plants and animals, an interest that Sauer himself came to share.

Sauer's approach to culture, however, owes most to the formulations of Alfred Kroeber. Kroeber was an eclectic scholar who had studied English at Columbia University before transferring to anthropology to work with Franz Boas, becoming his first successful PhD student in 1901 (Steward 1973). Kroeber was something of a polymath, also having a serious interest in psychoanalysis which he was eventually forced to drop in the interests of his anthropological studies. He continued to contribute to a broad range of scholarship throughout his life, including linguistics and archaeology as well as anthropology and psychology. His obvious disdain for disciplinary boundaries struck a sympathetic chord with Sauer. His conception of anthropology as the 'natural history of culture' was also one that Sauer would readily have approved, although Kroeber thought more in terms of the humanities than the natural sciences. For, despite Sauer's avowed interest in 'the aesthetic qualities of the landscape' (1925, p.48) and his affirmation of the need for a subjective approach to the understanding of such qualities, he remained wedded to a naturalistic philosophy of science (Entrikin 1984). By contrast, Kroeber's 'natural history of culture' was concerned with description and classification rather than with the search for causality or cultural origins. Arguably, the two major components of Kroeber's thought were his insistence on the *characterization* of the salient aspects of particular cultures and their accurate *classification*. Sauer's project was altogether more ambitious.

Kroeber was a prolific author who made a number of important theoretical contributions on the nature of culture (Kroeber 1944, 1952). At a comparatively late stage in his career he undertook a critical review of the concept and definition of culture, with Clyde Kluckhohn, which attempted both to clarify his own ideas and to summarize current thinking (Kroeber & Kluckhohn 1952). The study undertook the heroic task of reviewing 164 definitions of culture under such headings as: descriptive, historical, normative, psychological, structural, and genetic. However, the study was not as inconclusive as this profusion of definitions might suggest. The authors were able to report a reasonable degree of consensus among contemporary social scientists concerning several key elements of the culture concept:

> Culture consists of patterns, explicit and implicit, of and for behaviour acquired and transmitted by symbols, constituting the distinctive achievements of human groups, including their embodiments in artifacts; the essential core of culture consists of traditional (i.e., historically derived and selected) ideas and especially their attached values; culture systems may, on the one hand, be considered as products of action, on the other as conditioning elements of further action (ibid. p.357).

Several aspects of this definition are worth commenting on in relation to Sauer's use of the term 'culture'. Clearly, Sauer shared Kroeber's emphasis on patterns of culture and on its essentially acquired, transmitted or achieved nature, as opposed to its allegedly ascriptive qualities. Similarly, Sauer shared Kroeber's belief that culture was the property of human groups not individuals, and that it was embodied in custom and tradition. Sauer, however, put rather more emphasis than Kroeber on the artefactual quality of material culture as opposed to its symbolic forms. He also had relatively little time for the discussion of ideas and values except where they were directly expressed in the landscape. Kroeber's definition also had a profound effect on cultural geography through his insistence on the super-organic character of culture, an approach which Sauer adopted all too uncritically.

The super-organic approach to culture

Following the sociological theory of Herbert Spencer, Kroeber (1917) distinguished the super-organic level of social organization from the organic and inorganic levels. Put simply, a super-organic approach adopts the view that culture is an entity at a higher level than the individual, that it is governed by a logic of its own, and that it actively constrains human behaviour. Sauer himself adopted the same position concerning the supra-individual nature of culture. For Sauer, therefore, human geography had nothing to do with individuals, but only with human institutions or cultures (Sauer 1941, p.7). One of his students, and an eminent cultural geographer in his own right, articulates the super-organic position with even more force:

> We are describing a culture, not the individuals who participate in it. Obviously, a culture cannot exist without bodies and minds to flesh it out; but culture is also something both of and beyond the participating members. Its totality is palpably greater than the sum of its parts, for it is superorganic and supraindividual in nature, an entity with a structure, set of processes, and momentum of its own, though clearly not untouched by historical events and socioeconomic conditions (Zelinsky 1973a, pp.40–1).

In this definition, 'culture' is treated as an entity that individuals merely 'participate in' or 'flesh out'. Culture is 'touched by' historical and socio-economic forces, not generated by them. Nor is culture generated by human agency, responding instead to its own internal momentum. In each of these respects the super-organic approach to culture runs counter to the emphasis of much contemporary social theory. It has therefore come in for a good deal of criticism from contemporary geographers who are better versed than Sauer in social and cultural theory (e.g. Agnew & Duncan 1981, S. W. Williams 1983).

James Duncan, for example, has argued that the super-organic mode of explanation reifies culture, treating it as an entity with independent existence and causative powers (Duncan 1980). According to the super-organic approach, culture can be explained only in its own terms. It cannot be reduced to the actions of individuals or explained in terms of social forces other than those of culture itself. Culture responds to laws of its own. Adoption of a super-organic approach to culture therefore severely limits the questions that may be asked. Basing explanations in the transcendental realm of a supraindividual culture, cultural geographers have often failed to address the wider social context in which cultures are constituted and expressed. The convergence of social and cultural geography discussed by Duncan and others therefore demands the rejection of a reified view of culture and the adoption of a more sociological approach in its place.

Despite its many weaknesses, several generations of cultural geographers have continued to adhere uncritically to a super-organic view of culture. Symptomatic of this approach is the cultural geographer's almost obsessional interest in the physical or material elements of culture rather than in its more obviously social dimensions. This focus on culture-as-artefact has led to a voluminous literature on the geographical distribution of particular culture traits from log buildings to graveyards, barn styles to gasoline stations.[5] In contrast, much less consideration has been given to the non-material or symbolic qualities of culture or to other dimensions of the concept that cannot be 'read off' directly from the landscape. Ironically, in view of Sauer's commitment to interdisciplinary study, anthropologists made a firm break with the analysis of isolated culture traits as early as 1935 when Ruth Benedict advocated the analysis of *patterns of culture*, studied as articulated wholes (Benedict 1935), a view that had been championed even earlier in Britain by Malinowski's studies of the Trobriand Islands (1922).

Following Sauer, cultural geographers have adopted an unnecessarily truncated view of their subject, confined to mapping the distribution of culture traits in the landscape. In the introduction to their highly influential *Readings in cultural geography* (1962), Wagner and Mikesell perpetrated an extreme version of this disciplinary myopia:

> The cultural geographer is not concerned with explaining the inner workings of culture or with describing fully patterns of human behavior, even where they affect the land, but rather with assessing the technical potential of human communities for using and modifying their habitats (ibid. p.5).

Even if 'the inner workings of culture' are off-limits for geographers, the embargo on describing 'patterns of human behavior, even where they affect the land' is extraordinarily restrictive. Their definition of cultural geography as concerned only with 'the distribution in time and space of cultures and elements of culture' seems equally short-sighted. The definition of culture in such narrow disciplinary terms did not pass entirely without comment, however. Within a couple of years, for example, Harold Brookfield was urging his colleagues to transgress the 'human frontiers of geography' (Brookfield 1964), beyond a strict interest in what is directly observable in the landscape. And both Wagner and Mikesell have both, in their own ways, modified their earlier strictures, extending the scope of cultural geography beyond the self-imposed boundaries of their earlier work (Mikesell 1977, 1978, Wagner 1975).

Today, the Sauerian view of cultural geography is being extended in several new directions. Old questions about 'agricultural origins and dispersals', for example, are being asked in new ways by geographers and archaeologists (e.g. Harris 1981), while Sauer's concern for ecological balance in natural environments is being reinvigorated as geography generates its own brand of 'green politics' (e.g. Doughty 1981, Pepper 1984). The focus of this book, however, is on those new directions in cultural geography that are taking the subject closer to the social rather than the physical sciences. The remainder of this chapter will consider some of the new directions that cultural geography has taken as a result of recent developments in humanistic geography, leaving the radical encounter between cultural geography and historical materialism until Chapter 2.

Cultural geography and the new humanism

During the 1970s, several geographers began to reorient the subject away from the social sciences towards the humanities (Ley & Samuels 1978). They were concerned about the apparent denial of human agency in contemporary social science and looked to the humanities for a more sympathetic treatment of human individuality, subjectivity, and creativity. Within cultural geography this realignment included a reassessment of the relationship between geography and literature (Tuan 1978). Geographers had for many years used literature as a source of evidence about past landscapes and privileged the novelist's ability to capture the subjective qualities of place. But now they were enjoined to emulate the literary style of the great regional novelists (Meinig 1983). To date, however, little of this work has sought to challenge the concept of culture established by the Berkeley School. Most of the work in humanistic geography and literature (Pocock 1981) has shared an élitist view of culture and an obsessive interest in landscape. Only recently have geographers begun to show any interest in literary analysis and in other conceptions of geography besides traditional forms of landscape interpretation.[6]

Humanistic geography was novel in the extent to which it examined the philosophical premises of geographical inquiry (e.g. Ley 1981a), at a time when the discipline was still preoccupied with technique and method rather than with more fundamental questions of epistemology. But there were also continuities with the past and some geographers (e.g. Meinig 1979) sought to locate cultural and historical geography within this longer tradition of humane inquiry. This version of humanistic geography, associated with the names of Clarence Glacken, David Lowenthal and Yi-Fu Tuan, among others, was focused on landscape, on ideas of Nature and on human consciousness itself. Two geographical mavericks, W. G. Hoskins and J. B. Jackson, also have a respected place in this tradition, although there are vital contrasts to be drawn between them. Hoskins is a pronounced anti-modernist who declared his distaste for contemporary landscapes in the final chapter of his classic book, *The making of the English landscape*:

> The industrial revolution and the creation of parks around the country houses have taken us down to the later years of the nineteenth century. Since that time, and especially since the year 1914, every single change in the English landscape has either uglified it or destroyed its meaning, or both (Hoskins 1955, p.231).

Jackson shares Hoskins' fascination with landscape, having founded and edited *Landscape* magazine for 17 years. But, unlike Hoskins, he has been most animated by 20th-century American landscapes and, in particular, by the human geography of the American South-West. His tastes are extremely wide-ranging and include vernacular landscapes, reflecting popular tastes, as well as élite landscapes, established by political authority (Jackson 1984). The breadth of his interests can be judged from some of his own contributions to *Landscape* magazine, which range from essays on mobile homes, trailer parks, and motels to commentaries on 'The abstract world of the Hot-Rodder' (1957-8) and 'The Domestication of the Garage' (1976). Jackson's interest in popular culture has not been widely shared within North American cultural geography. But his devotion to regional geography, and his concern for the quality of geographical writing, have received much wider acclaim.

Indeed, regional geography has experienced something of a revival lately with an infusion of energies from diverse sources (Pudup 1988). The case for regional description as 'the highest form of the geographer's art' has been confidently reasserted by traditionalists such as Hart (1982), while theoretically more sophisticated versions of regional geography have also been proposed (Gilbert 1988). But what exactly does this revival involve, and how have these geographers confronted the age-old problem of geographical description (Darby 1962)? Recent work in geography and literature illustrates both the problems and the potential. It might have been thought, for example, that there was nothing left to be said about the geography of Hardy's Wessex after Darby's (1948) pioneering essay and Birch's (1981) exhaustive researches. But John Barrell's (1982) analysis of the many 'geographies' in *Tess of the d'Urbervilles* (1891) and *The return of the native* (1878) has a richness and subtlety that puts his geographical predecessors to shame. Rather than trying to identify 'real' places in the Wessex novels or attempting to write a 'regional geography' of Hardy's Wessex, Barrell examines how localities and spaces in Hardy's fiction are constructed and mapped out by the characters in the course of the novels, and how he manipulates the narrative to reveal to the reader a range of different subjective geographies.

It is no coincidence that the geographer's interest in literature began with Thomas Hardy and Walter Scott, later extending to Arnold Bennett and the other 'regional novelists' (Darby 1948, Paterson 1965, Gilbert 1972, Hudson 1982). These authors had an obvious appeal because of the incorporation of 'real' places in their novels, transforming them through their imagination into creative 'geographies of the mind' (Lowenthal & Bowden 1976). Geographers have, however, shown little interest in the symbolic representation of place in literature or in what those representations tell us about the non-literary world. Why, for example, are James Bond movies always set in such obviously 'exotic' locations, and how do science fiction novels, like all utopian works, project contemporary social relations on to the imagined geographies of the future? Our experience of place is now thoroughly mediated by what we read and what we see on television, yet the media have only recently received any serious treatment from geographers (Burgess & Gold 1985).

The most serious weakness of humanistic studies of literature, however, is their shallow treatment of social context. As one sympathetic critic has argued in a general assessment of the problems of cultural and humanistic geography:

> In retrieving man [*sic*] from virtual oblivion in positivist science, humanists have tended to celebrate the restoration perhaps too much. As a result values, meanings, consciousness, creativity, and reflection may well have been overstated, while context, constraint, and social stratification have been under-developed (Ley 1981b, p.252).

Various alternative approaches to the geographical study of literature have begun to emerge from the radical critique of humanistic geography, although geographical analyses of literature from a materialist perspective remain the exception (Silk 1984). These approaches are considered at greater length as part of a wider discussion of cultural materialism in Chapter 2.

Conclusion

This chapter has attempted to demonstrate the urgent need for a revised conception of culture in geography. However laudable the achievements of the Berkeley School, its domination of cultural geography effectively prevented a whole range of alternative approaches from being considered. The deficiencies of a super-organic approach are now generally recognized and a more active conception of culture is required, acknowledging the extent to which cultures are humanly constituted through specific social practices. An exclusive interest in the physical expression of culture in the landscape now also seems unnecessarily restrictive. Élitist concepts of culture, concerned only with the Great Tradition of English literature and the fine arts is, likewise, no longer acceptable. Instead, cultural geographers are beginning to recognize a plurality of cultures and to shift attention away from the analysis of a few privileged texts towards an analysis of the social relations through which cultures are produced and re-produced.

This reorientation of cultural geography involves a convergence of interests with social geography and an openness to developments in the broad, interdisciplinary field of cultural studies. For those who take a narrow view of what is the proper subject of geographical inquiry, this will take them beyond geography altogether. But in resolutely ignoring the developments in cultural studies that have preoccupied other social scientists in recent years, geographers have left themselves in a position of extreme isolation that accounts for much of the current defensiveness about disciplinary boundaries.[7]

This book places the geography of culture squarely within the social sciences. The literature it reviews is not confined to the study of landscape or environment as these terms are conventionally defined. Nor is it concerned only with specific places, except insofar as the processes described have a distinctive geography. The conception of cultural geography advanced here focuses on the way cultures are produced and reproduced through actual social practices that take place in historically contingent and geographically specific contexts. It is in the specification of context in its fullest sense that geography can make its most immediate contribution to cultural studies, rather than through a unique interest in the geographical distribution of 'cultural traits'. Geographers can no longer take the definition of culture as given; it is a contested term the meaning of which must now be considered problematic.

Notes

1 There is an extensive literature on Carl Sauer and 'Sauerology' (Mikesell 1986). In writing this chapter, the following sources have been most useful: Entrikin (1984), Leighly (1976), Parsons (1976, 1979), and M. Williams

(1983). The best introduction to Sauer's voluminous writings is through the collection of his essays, edited by Leighly (1967).

2 Sauer claimed to have no interest in these *ex cathedra* pronouncements. He was, though, according to Entrikin (1984), 'a philosopher in spite of himself'.

3 While firmly attached to the landscape tradition, Sauer's work on agricultural origins and dispersals (e.g. Sauer 1952) also prefigured later developments in spatial analysis concerning the diffusion of innovations (e.g. Hägerstrand 1967). Hägerstrand's main concern, however, was with innovation as a general (spatial) process rather than with visible aspects of the cultural landscape of specific geographic areas. Hägerstrand's monograph on innovation diffusion, originally published in Swedish in 1953, did not cite Sauer's work, which was incompatible with the methodological individualism that characterized Hägerstrand's approach. A recent critique of diffusion theory (Gregory 1985) charges Hägerstrand with neglecting the structures of social relations and systems of social practices that govern the diffusion of innovations, a criticism which applies equally to Sauer.

4 That Sauer's students referred to him as the 'Old Man' while he was still only in his 40s is indicative of his pedagogic style. That former students still refer to him by this title suggests that their recollection of 'the later Sauer years' (Parsons 1979) is tinged with nostalgia.

5 Examples include Terry Jordan's studies of Texas log buildings and graveyards (Jordan 1978, 1982) or, from the perspective of American popular culture, the encyclopaedic approach of the Society for the North American Cultural Survey (Rooney *et al.*, 1982). All too often, the rationale for these studies is unclear. An exception is Fred Kniffen's discussion of the origins of his interest in folk housing (Kniffen 1979) which centres on the human transformation of raw materials, giving the study of log buildings a strategic significance in demarcating the boundary between 'nature' and 'culture'.

6 Even Duncan and Duncan's critical engagement with literary theory is described as a (re-)reading of the landscape (Duncan & Duncan 1988).

7 A recent issue of the *Transactions of the Institute of British Geographers* carried half a dozen papers on 'the unity of geography' (Freeman 1986). The crisis in American geography (Haigh 1982) has also produced a strange blend of nervous introspection and public protestation of which 'National Geography Awareness Week' was only the most striking example.

Chapter two
Problems and alternatives

Despite its obvious deficiencies, geographers have been slow to challenge the uncritical conception of culture inherited from Carl Sauer and the Berkeley School. Indeed, many geographers do not seem to regard culture as at all germane to their work, ignoring it altogether or treating it as a trivial residue left unexplained in more powerful analyses of political economy. In other cases, cultural variation has been regarded as an unfortunate cause of minor anomalies in the application of general models.[1] In general, cultural geography has remained untouched by the theoretical ferment that has been taking place elsewhere in cultural studies. Raymond Williams, Stuart Hall, Clifford Geertz, and Claude Lévi-Strauss have, among others, all adopted positions that imply a fundamental challenge to the basic assumptions of cultural geography. But, with only rare exceptions, geographers have remained singularly aloof from this intellectual turmoil. Nor can they adopt a *laissez faire* attitude towards the meaning of culture in the mistaken belief that the issue has been satisfactorily resolved elsewhere in the social sciences. For 'the most penetrating criticisms of traditional cultural geography have centered on its neglect of issues examined habitually by [other] social scientists' (Mikesell 1978, p.10), including questions of power, inequality, gender, class, and race.

Though geographers have been slow to take an interest in the interdisciplinary field of cultural studies, the opportunity is not yet lost. Cultural studies is not a closed book. It is an open and vigorous field that invites the active participation of social and cultural geographers. To enter this field, however, geographers must be prepared to interrogate the culture concept in a much more critical way, rather than passively accept the received wisdom of the Berkeley School. An appropriate starting point for a more interdisciplinary approach is to re-evaluate the development of cultural theory in 19th-century anthropology. From there, it will be possible to trace the growth of a more critical line of thinking, described here as a form of cultural materialism.

The study of culture and 19th-century anthropology

While North American cultural geographers were influenced most strongly by Franz Boas and his students, the equivalent influence in Britain was exercised by the English anthropologist, E. B. Tylor (1832-1903). Tylor was keeper of the Pitt-Rivers Museum in Oxford, where he was also Professor of Anthropology. In *Primitive culture* (1871) he advanced a definition of culture that is still regarded as a bench-mark in the development of anthropological theory. Tylor defined culture as 'that most complex whole which includes knowledge, belief, art, morals, law, custom, and any other capabilities and habits acquired by man (*sic*) as a member of society' (Tylor 1871, p. 1). At first glance, the definition has much to recommend it. It is holistic and inclusive, acknowledging both the breadth and complexity of culture together with its fundamentally social nature. Its strengths are, however, also its weaknesses. It is too inclusive to serve much analytical purpose, not providing any means for distinguishing the cultural from other elements of society, such as the political or the economic. Nor does it suggest appropriate ways of handling the complexity of forms in which culture is expressed.

Why, then, has Tylor's statement been considered definitive? The answer lies partly in its strategic importance in the history of anthropology, providing a sense of unique professional identity to an emerging discipline. In many ways, Tylor personified the development of the discipline itself. He was both a child of his times and the herald of a new generation. On the one hand, his work was thoroughly infused with Darwinian ideas about evolutionary change and human development. He was also a keen advocate of the need for an empirical 'science of man' to replace the ungrounded speculations of theologians and moralists. Yet, on the other hand, Tylor was distinguished from his peers in being the first British anthropologist to regard culture as a central, problematic focus of scholarly investigation.

Tylor set himself the goal of making the beliefs and actions of 'primitive man' seem less strange by making them intelligible to the Victorians' conception of rational 'scientific man'. Thus, 'primitive religion' was to be understood as the 'natural' beliefs of people whose mental powers were similar to those of the Victorians except that they were hampered by a lack of proper (scientific) information. In the case of culture, Tylor's evolutionary thinking prompted similar conclusions: 'primitive' culture was not intrinsically different from that of Victorian England. Indeed, it was argued, traces of earlier cultures continued as 'survivals' into the present. Tylor's approach was that of the antiquarian, using the comparative method to point out differences between cultures at different 'stages' of technological development. (Similar ideas on the origins of patriarchy and the evolution of the family were being propagated at the same time by Lewis Henry Morgan in *Ancient society* (1877) and by Friedrich Engels in *The origin of the family, private property and the state* (1884).)

Tylor's adoption of the comparative method had other advantages, which have been described as the 'ulterior motives' of his anthropology (Burrow 1966, p.251). The method allowed Tylor to render 'primitive' people acceptable as ancestors by making their beliefs and practices comprehensible rather than merely quaint, exotic, or shocking. It also provided him with a means of criticizing contemporary 'superstitions' where he could show that these were related to former cultural traditions which 'scientific evidence' now rendered untenable. For all its faults, then, Tylor's anthropology was a determined attempt to regard culture critically.

If Tylor's contribution was to introduce a more critical approach to culture among his 19th-century anthropological colleagues, the contemporary problematic can best be described as the critique of 'culturalism' – a spurious belief in the explanatory power of culture, abstracted from its material circumstances.[2]

The problem of culturalism

Unlike Tylor's critical speculations on the development of culture, culturalism refers to an unreflective, common-sense approach. It assumes that culture is a self-evident and unproblematic category that can be used to explain people's behaviour. Culture is given causal powers, and people are said to do things *because of* their culture. A culturalist point of view seeks to explain ideas and practices with respect to culture, rather than seeing culture as something to be explained. Two examples illustrate the problems of culturalist analyses and indicate the need for more critical alternatives.

On 12 May 1984, the *Los Angeles Times* reported that a high-ranking official of the federal Department of Housing and Urban Development had sought to explain the poor living conditions of the Hispanic population in terms of their cultural 'preference' for overcrowded housing. Rather than citing poverty as the explanation for their overcrowded condition, or their uncertain political status as 'undocumented workers', the official chose to emphasize their large family size and 'preference' for living with their extended families. He failed to recognize that 'preferences' are subject to material constraint and that it is unreasonable to infer 'preferences' directly from behaviour. Neither can the statement be rejected simply as a matter of individual ignorance. It is a reflection of institutionalized attitudes and structured inequality. The implication that Hispanic 'culture' explains their poor living conditions, regardless of their material circumstances, is an example of a culturalist explanation.[3]

The second example also shows how culturalist analyses can have extremely damaging social consequences. It is taken from a confidential report on the Rastafarian movement in New York, prepared by the Intelligence Division of the New York City Police Department and leaked to the *Caribbean Review* (New York City Police Department 1985). The police sought to justify the report not simply in terms of the growing number of Rastafarians living in New York (estimated by the police department at 10000), but also in terms of the alleged association between Rastafarianism and crime. The report begins with the disingenuous disclaimer that 'all Rastafarians are not criminals'. It repeatedly makes the distinction between the 'true' Rastafarian who is a 'law-abiding, proud individual' and the 'criminal element' that is responsible for the criminal reputation of the group. These distinctions are made as statements of self-evident fact for which there is apparently no need to offer empirical support. In other respects, however, the report is based on the assumption that the police's Intelligence Division has privileged knowledge of New York's Rastafarians and that it is charged with the self-appointed task of 'translating' the group's arcane practices and criminal argot for the benefit of the wider community. This sometimes takes a very direct (and often quite ludicrous) form of literal translations of Rastafarian speech into the (no less exotic) language of the police department.

Equally characteristic of culturalist analyses is the recourse to a pseudo-scientific idiom in an effort to lend credibility to what is actually hearsay if not quite imaginary. The New York police maintain that 'massive amounts of ganja are smoked daily by Rastafarians' although, they continue, 'it has yet to be established scientifically that massive amounts of marijuana consumed daily can alter mental attitudes'. Nonetheless, they are happy to speculate that heavy ingestion of marijuana is largely responsible for the Rastafarians' stoic attitude towards violence', even though they admit that the Rastafarians would themselves deny this.[4] There is no need to labour the point. In the space of a short report, the New York City Police Department can attribute a whole range of undesirable characteristics to Rastafarianism, reaching alarmist conclusions about this 'very rapidly growing cult' and its 'potential for manipulation'. Despite the report's insistence that only a minority of the Rastafarians are engaged in criminal activities, it imputes a connection between a particular group of people and the perceived threat of criminal violence. The use of culturalist explanations casts a disreputable aura around the whole Rastafarian movement on the basis of mere allegation.

These examples may seem easy targets for criticism but the form of argument they employ is not limited to government agencies or police officials. Social scientists have themselves contributed to the perpetuation of culturalist ideas with equally damaging consequences. The following sections pursue these ideas, demonstrating the political effects of culturalist arguments about the city and the 'culture of poverty'.

Culturalism and the city

Until comparatively recently, most urban researchers took a highly uncritical attitude towards culture. Manuel Castells, for example, condemned the entire corpus of the 'Chicago School' of urban sociology for its culturalist values, taking particular exception to Louis Wirth's classic essay on 'Urbanism as a way of life' (1938) in which Wirth defined urbanism in terms of the size, density and social heterogeneity of urban settlements. Though Castells accepts that the distinction Wirth draws between primary and secondary social ties (face-to-face versus impersonal transactions) may have emerged first in cities, he denies the existence of a specifically urban culture: 'everything described by Wirth as "urbanism" is in fact the cultural expression of capitalist industrialization, the emergence of the market economy and the process of rationalization of modern society' (Castells 1976, p.38). Unlike Wirth, Castells refuses to abstract urban culture from its foundations in the material world. His criticisms, subsequently elaborated in *The urban question* (Castells 1977), represented one of the first alternatives to Chicago School sociology, provoking a series of debates about the significance of space in urban social theory (Smith 1980, Saunders 1981).

David Harvey also launched an attack on culturalism in *Social justice and the city* (1973), arguing that 'the main thrust of the Chicago school was necessarily descriptive' and finding Engels' approach to urbanism 'more consistent with hard economic and social realities than . . . the essentially cultural approach of Park and Burgess' (ibid. pp. 131-3). Like Castells, Harvey sought to replace the Chicago School's ecological analysis with a materialist analysis of the city's political economy. And while there were important differences in the theoretical approaches they adopted and in the empirical work they undertook, both Harvey and Castells placed considerably greater emphasis on the problematic status of the urban in their critique of 'urban culture' than on the nature of culture itself. Harvey has since undertaken detailed historical research on 19th-century Paris in an attempt to bring theory and experience closer together, showing how the 'urbanization of capital' was paralleled by the 'urbanization of consciousness' (Harvey 1985a, 1985b). Castells has also completed a programme of empirical research, aimed at producing a cross-cultural theory of urban social movements (Castells 1983) and, in its own way, addressing questions of cultural theory. Studies of urban poverty have so far failed to make much progress in breaking loose from the culturalist assumptions of the 'culture of poverty' thesis first advanced in the 1960s.

Culturalism and poverty

The 'culture of poverty' is a term that was coined by Oscar Lewis, an American anthropologist working principally among low income groups in Mexico, Puerto Rico, and New York during the 1950s and 1960s. His ideas reached a large audience through an influential article in *Scientific American* (1966) and, at much greater length, through a series of semi-autobiographical life-histories recorded by Lewis in the field (Lewis 1959, 1961, 1964, 1965). The popular appeal of these books stems from Lewis' interest in prostitution, crime and gambling as well as in more mundane anthropological aspects of family life among the poor. But it was the concept of a 'culture of poverty' that stimulated the most heated controversy among professional social scientists (Valentine 1968, Leacock 1971) and which had the most unfortunate political consequences. The controversy revealed many of the inadequacies of culturalism in general.

Lewis defined the 'culture of poverty' as a way of life shared by poor people in given historical and social contexts. Its characteristics included:

(a) a lack of effective participation in the major institutions of the larger society;
(b) minimum organization beyond the nuclear and extended family;
(c) absence of childhood as a prolonged and protected stage in the life cycle; early initiation into sex; free unions or consensual marriages; and a trend towards female- or mother-centred families; and

(d) strong individual feelings of marginality, helplessness, dependence, and inferiority (Lewis 1965, pp.xlv-xlviii).

In general, Lewis maintained, the 'culture of poverty' was 'a relatively thin culture' with a great deal of pathos, suffering and emptiness. Indeed, 'the poverty of culture is one of the crucial aspects of the culture of poverty' (ibid. p.xlvii).

The principal deficiency of Lewis' formulation, apart from its inherent middle-class bias, is its failure to treat the culture of poor people as a positive response to their low economic status and social subordination. Instead, Lewis regards the 'culture of poverty' as a closed system, independent of wider socio-economic forces. After all, if people do not participate in the 'major institutions' of society, it may be because they are excluded rather than because they refuse to be integrated with the rest of society. The preponderance of single-parent families and female-headed households, which Lewis regards as inherently pathological, may similarly be a reflection of institutional forces, such as the current administration of social services.[5] In as much as these wider forces impinge on the family patterns described by Lewis as a 'culture of poverty', it is highly dubious to imply that they are elements of a voluntary set of cultural preferences. Much of the 'culture of poverty' literature can, in fact, be read as an instance of the reactionary habit of 'blaming the victim' for problems which are not of their own making and which they have little power to alter.

These criticisms are all the more telling in the light of Lewis' insistence that the 'culture of poverty' is self-perpetuating from generation to generation because of its debilitating effect on 'slum children' who, by the time they are six or seven, 'have usually absorbed the basic values and attitudes of their subculture and are not psychologically geared to take full advantage of changing conditions or increased opportunities which may occur in their lifetime' (Lewis 1965, p.xlv). This statement embodies many of the weaknesses of the culturalist approach and helps explain why its critics find it so objectionable. Not only is culture treated as a fixed and static entity that is 'absorbed' in a once-and-for-all process of childhood socialization, but the problems of 'slum children' are attributed to their 'psychological orientation' rather than to the material conditions in which they live. A more critical concept of culture would challenge all these assumptions. For even under conditions of extreme poverty, cultures are not passively received but actively forged.

None of these criticisms would be of more than academic interest, however, were it not for the fact that social policy has been based on ideas that derive from Lewis' 'culture of poverty' thesis. The controversial Moynihan report on 'The Negro family' (1965) is the most obvious example, adopting the language and many of the assumptions of the 'culture of poverty'. The report suggested that the problems of poor black people were the result of a pathological and self-perpetuating subculture. Moynihan was quite explicit about this, arguing that 'the deterioration of the Negro family' is 'the fundamental source of weakness of the Negro community at the present time' (quoted in Rainwater & Yancey 1967, p.51).

Despite the liberal intentions of the report and its call for national action, its unintended consequence was to direct federal attention away from the root causes of black people's problems – the structured inequalities of power that divide contemporary American society along lines of race, class and gender. By concluding that 'the Negro family' was itself to blame for the poverty and deprivation of black people, the report absolved the rest of society from its responsibility for dealing with the effects of past injustices and present-day inequalities. As race-specific policies again come under scrutiny in the America of the 1980s, it is not surprising to find contemporary social policy once again producing analyses that resemble Lewis' 'culture of poverty', even if they are now expressed in a slightly different language.

William Julius Wilson's *The truly disadvantaged* (1987) is ostensibly a liberal response to the problems of the inner city 'underclass'. Although Wilson argues that the social problems of urban life in the United States are, in large measure, the problems of racial inequality (ibid. p.20), he does not accept that the current rise of 'social dislocations' among members of the ghetto underclass are due mainly to contemporary racism (ibid. p.10). He criticizes conservative thinkers for their uncritical adoption of Lewis' 'culture of poverty' thesis but commends earlier writers on the history of the black ghetto who referred to ghetto culture as a 'self-perpetuating pathology' (ibid. p.4). Elsewhere, too, Wilson describes the 'vicious cycle' of poverty, perpetuated through the family, community and schools (p.57). Like Lewis, Wilson regards out-of-wedlock births, single-parent families and female-headed households as *a priori* evidence of 'social dislocation', part of a 'tangle of poverty' in the inner city. He traces the problem to a general absence of 'mainstream role models' (ibid. p.56) and, specifically, to the absence of a sufficiently large 'marriageable pool' of black men in steady employment who are in a position to support a family (ibid. p.83). It would be easy to reject similar arguments from a white social scientist as reactionary, even as racist. Written by one of America's leading black sociologists, however, the argument is much harder to dismiss. Wilson's key theoretical concept is 'social isolation' rather than the 'culture of poverty' (ibid. p.61). In his desire to help 'the truly disadvantaged' through universal programmes that enjoy widespread support from a broad constituency, Wilson runs the risk of perpetuating a culturalist analysis that ignores the very structures of inequality and power he seeks to address.

As the above examples aim to demonstrate, there are very practical reasons, as well as purely academic ones, for challenging culturalist assumptions. Political arguments are often couched in cultural terms and cultural arguments are rarely free of political consequence. Because of the weaknesses of culturalist modes of explanation, alternative perspectives must be explored that give greater credence to the material dimensions of economy and society. Most such approaches risk simply reducing 'culture' to 'society', or advancing tautologous arguments about the 'relative autonomy' of culture. One of the most appealing alternatives, that avoids both of these pitfalls, is the cultural theory of Raymond Williams who describes his approach as a form of cultural materialism.

Cultural materialism

Cultural materialism can be defined as a particular application of the Marxist method of historical materialism to the field of cultural studies.[6] The common emphasis in all materialist analyses is their refusal to treat the realm of ideas, attitudes, perceptions and values as independent of the forces and relations of production. Instead, culture is seen as a reflection of the material conditions of existence. A materialist approach, by definition, concentrates on the material or economic basis of human society. The question is, of course, whether this is a legitimate procedure for cultural analysis where creativity and individuality are often thought to be beyond the scope of such mundane influences. Is it possible to produce a materialist analysis of culture that does not simply become an argument in economic determinism? Williams' work was a lifelong exploration of this fundamental question.

Raymond Williams (Fig. 2.1) was born in the Welsh town of Pandy in Gwent, not far from the English border. He grew up with a strong sense of the distinct identity of Wales within the British Isles, reflected in his first novel *Border country* (1960). The son of working-class parents, he was extremely conscious of his anomalous position within the class structure of British academia from the time he went up to Cambridge as an undergraduate in 1939. Williams was active in student politics and was, for a short while, a member of the Communist Party. His formal education was interrupted by World War II, in which he served as a captain in an anti-tank division.[7] Williams then worked as a tutor in adult education before being elected a Fellow of Jesus College, Cambridge, in 1961. Officially a Professor of Drama, his interests were

Figure 2.1 Raymond Williams, 1921–1987

extremely wide-ranging, encompassing literary criticism, media studies, and political analysis.

Williams' theoretical ideas were worked out most thoroughly in *Marxism and literature* (1977) and applied historically in *Culture and society* (1958) and in its sequel, *The long revolution* (1961). One of the central questions that Williams addresses in these works is how to construct a materialist analysis of culture that does not reduce to a simple distinction between 'base' and 'superstructure', the latter always being determined by the former. The question of 'structural determination' is a crucial one in Marxist theory. For several critics (e.g. Duncan

& Ley 1982), this has been a central aspect of their rejection of 'structural Marxism' which they interpret as giving only a subordinate place to culture as the mere reflection of economic forces. Whether or not one accepts these criticisms (which seem only to apply to the most economistic forms of Marxism), Williams' attempt to grapple with these central issues is greatly to be welcomed.

As Williams argues in *Marxism and literature*:

> it is not 'the base' and 'the superstructure' that need to be studied, but specific and indissoluble real processes, within which the decisive relationship, from a Marxist point of view, is that expressed by the complex idea of 'determination' (1977, p.82).

Characteristically, Williams points out the linguistic complexities of the word 'determine' in Marx's work. The usual word that Marx employed (*bestimmen*) refers to the process of setting limits. But Williams also identifies a 'scientific' usage of the term, as in the phrase 'determinate conditions', which refers to a set of definite, relatively fixed conditions or circumstances. Change can then be understood in terms of altered conditions or combinations of circumstance that can be explained, if not actually predicted. Marx used the concept of determination in several key passages of his work. In the *Critique of political economy* (1859), for example, he argued that the mode of production in material life determines the general character of the social, political, and spiritual processes of life, while in *The German ideology* (1846) he and Engels asserted that life was not determined by consciousness but consciousness by life. Different translations give rise to very different kinds of Marxism, from a deterministic reading of the 'iron laws' of economic history to a more active interpretation of the process of class struggle under various 'determinate conditions'. Williams attempts to resolve these issues by quoting from one of Engels' letters where he wrote that: 'We make our history ourselves, but, in the first place, under very definite assumptions and conditions' (Williams 1977, p.85). The idea of 'determination' is here returned to its original meaning as 'the setting of limits', effectively restoring an active conception of human agency but one which is subject to 'very definite conditions'.

By focusing on 'specific and indissoluble real processes', Williams offers a view of determination that is thoroughly appropriate to a reconstituted cultural geography. Rather than seeing determination as something that takes place in relation to a static mode of production, he adopts a more active, conscious view of historical experience, recognizing multiple forces of determination, structured in particular historical situations (ibid. p.88). Having clarified some of the conceptual baggage of Williams' cultural materialism, it remains to be seen how his ideas can be applied in practice.

Culture and society

Much of Williams' work can be understood in terms of an intellectual convergence between literature and sociology, particularly English literature and Marxist sociology. His works begin characteristically with a careful explication of the derivation of certain 'keywords' – such as culture itself – and proceed with a highly developed sensitivity to social and historical context (Williams 1976). For example, *Marxism and literature* (Williams 1977) begins with a discussion of the quartet of concepts which are basic to Williams' subsequent discussion of cultural and literary theory: culture, language, literature, and ideology. He proceeds to show how the contemporary meaning of 'culture' originated with the German Romantics; how it became a synonym for 'civilization' in 18th-century English usage; and how it only took on its contemporary meaning ('a whole way of life') with the Industrial Revolution. As Williams elaborates in *Culture and society* (1958), these changes were not accidental but rather were related to the evolution of the meaning of other 'key words' (such as industry, democracy, class, and art) all of which owe their contemporary meaning to the same formative period of intense social change.

Williams' work comprises a thorough attack on the kind of culturalism described earlier in this chapter. His criticism of culturalism (and of idealism in general) centres on its tendency to ignore the complex social relations that lie behind the production of culture. Indeed, a central tenet of Raymond Williams' cultural materialism is the notion that cultural forms of all kinds are the result of specific processes of production. This applies as much to literature and art as to the world of advertising, television, and the other communications media where 'cultural production' has a more obvious relevance (Williams 1962, 1974). Two brief examples illustrate the application of cultural materialism in Williams' work.

The first concerns the Pre-Raphaelite Brotherhood whose explicit rejection of academic conventions and overt dedication to observing Nature at first hand generated considerable popular interest in the second half of the 19th century. Williams rejects the idealist interpretation of the Pre-Raphaelite Brotherhood which focuses on individual biographies and aesthetic considerations of personal taste and style. He shows instead how the success of the Pre-Raphaelites was related to a particular set of material circumstances. Specifically, Williams shows that the social origins of the Brotherhood were rooted in the commercial bourgeoisie. He demonstrates how the Pre-Raphaelites deliberately set themselves against the main cultural tendencies of their period and class, and yet how they were able to attract patrons from this same commercial, bourgeois, and generally provincial class (Williams 1981, pp.77-8). The stated principles and stylistic conventions of the movement are set firmly in their social and historical context, unlike idealist or culturalist analyses which imply that tastes and styles are independent and free floating.

Williams' materialist analysis extends to the content and style of the artists' work, setting their nostalgic 'medievalism' against the pervading 'industrialism' of the period, giving rise to a type of naturalism and a decorative kind of beauty that was peculiarly acceptable to their bourgeois patrons. Both artists and patrons are situated by Williams in the context of the newly emergent class structure of the latter part of the 19th century. Significantly, moreover, though several members of the Brotherhood outwardly adopted unconventional, 'bohemian' life-styles, only in the case of William Morris were the commercial practices of capitalism itself directly challenged. And, even in his case, there were several unresolved contradictions between his socialist politics, his aesthetic ideals and the commercial realities of producing for sale.[8]

The second example also concerns a 19th-century artistic movement which Williams interprets against the historical background of a materially changing social order. The Bloomsbury Group were a much looser type of cultural formation of writers, artists, and academics, including Virginia Woolf, Clive and Vanessa Bell, Lytton Strachey, and John Maynard Keynes. Yet they too shared common social origins, coming from professional and administrative families. Educationally they represented the product of the newly reformed 'public' school and university system. Several of the leading members of the Group had known each other at Cambridge and later lived near one another in the area of London from which they took their collective name. Despite their social origins among the ruling class, the Group's educational background and cultural interests clearly set them against the industrial and commercial ethos of their period and class. Williams, therefore, sees the Group as a 'fraction' of this class, self-consciously contrasting their literary achievements and academic skills with the perceived stupidity, incompetence, and prejudice of the rest of their class who wielded political and economic power through Parliament, the City and the Civil Service. The Group owed much of its unity to members' shared antagonism towards unmanaged capitalism, towards militarism, and colonialism, and towards the subjugation of women and the denial of sexual freedom (Williams 1980, pp.148-69; 1981, pp.79-81).

In both examples, Williams relates the cultural to the social but does not simply read off culture from a simplified social and economic history. He is highly critical of mechanistic arguments which are forced and superficial. Williams' version of cultural materialism manages to avoid reductionism by relating literary and artistic production to a whole way of life rather than to the economic system alone. This is clear from Williams' definition of culture as 'a realised signifying system': a set of signs and symbols that are embedded in a whole range of activities, relations, and institutions, only some of which are manifestly 'cultural', others being overtly economic, political, or generational (Williams 1981, pp.207-9). In all his case studies, the cultural is interpreted in terms of a wider set of relations that include aesthetics and morals as well as economics and politics. For Williams, culture is 'a whole way of life', a general social process, not confined to the intellect or the imagination (Williams 1958).

Though these examples illustrate the method of cultural materialism and its ability to reveal the relationship between culture and society, little has yet been said about the grounding of Williams' work in particular places, or the model he offers for a materialist cultural geography. These ideas are developed most clearly in *The country and the city* (1973) which, because of its subject matter, is probably the best known of Williams' works among geographers. It consists of a series of literary interpretations through which Williams traces the persistent way in which 'country' and 'city' have evoked simultaneously positive and negative feelings during several centuries of often quite rapid social change. The country is loved for its 'peace, innocence, and simple virtue', but scorned for its 'backwardness, ignorance [and] limitation'. The city is esteemed as 'an achieved centre: of learning, communication, light', but loathed as 'a place of noise, worldliness and ambition' (ibid. p.1). Not only does Williams show the ambiguities and contradictions in people's attitudes to the country and the city; he also reveals how these ideas are rooted in the *actual material connections* between country and city. Whether the subject is the English country house or Hogarth's *Gin lane*, Williams reveals how contemporary attitudes to country and city reflected changing socio-economic conditions and, in particular, how pastoral visions of the countryside concealed the labour that made the rural idyll possible for the few by denying basic rights to the many.

Some readers of Williams' work are disturbed by the apparent contradiction between his radical, socialist politics and his analysis of what are almost invariably élite sources – the great literature of the age. Nor can the contradiction be resolved simply in terms of the available historical evidence which is certainly biased in favour of élite sources, but not to the exclusion of other material that more clearly reflects contemporary popular culture. This problematical emphasis in Williams' work has been defended as an attempt to provide a critique of the patterns of established culture 'at their strong points' rather than concentrating on what may have been easier targets (Blackburn 1988, p.14). And certainly Williams' use of his sources is consistently radical even if their origin is not. A further criticism is Williams' concentration on literary material, despite his awareness of the power of visual material, explored in other of his books (e.g. Williams 1962, 1974). Here, too, he can be defended. Although he concentrated on literary material in *The country and the city*, similar work on landscape painting by other authors has amply confirmed his argument. John Barrell's *The dark side of the landscape* (1980) is an outstanding example of this genre, analyzing the depiction of the rural poor in 17th- and 18th-century English painting against a background of changing social relations during the period. Similar work is now being undertaken by geographers, exploring the relationship between art and agrarian change (Prince 1988), or, more generally, between social formation and landscape symbolism (Cosgrove 1985b).

In order to understand such persistent and complex feelings as those evoked by the symbolic opposition of 'country' and 'city', Williams employed his notion of a 'structure of feeling' (Williams 1977, pp.128-35). It is an important concept for cultural geographers, sharing something of the meaning of 'sense of place' but going well beyond it in several respects. By 'structures of feeling', Williams attempted to identify 'the particular quality of social experience and relationship, historically distinct from other particular qualities, which gives the sense of a generation or of a period' (ibid. p.131). The concept refers to meanings and values as they are actually lived, not just to formal worldviews or ideologies. It refers to present and future, as well as to past, experiences, and to such intangible qualities as 'characteristic elements of impulse, restraint, and tone' as well as 'specifically affective elements of consciousness and relationships' (ibid. p.132). These feelings constitute a 'structure' in the sense that they are a 'living and inter-relating continuity', a set, with specific internal relations, interlocking and in tension. There is more than a passing similarity, then, with Bourdieu's (1977) concept of 'habitus' which refers to the cognitive structure of any social group, comprising the sedimented history of particular practices that arise to meet certain objective conditions and which thereby serve to reproduce these conditions.[9] Given Williams' interests, however, his own concept has a special relevance to the analysis of art and literature, in the sense that form and convention provide one of the most significant clues to the recognition of a particular 'structure of feeling'.

A final example from Williams' writings, concerning the work of the

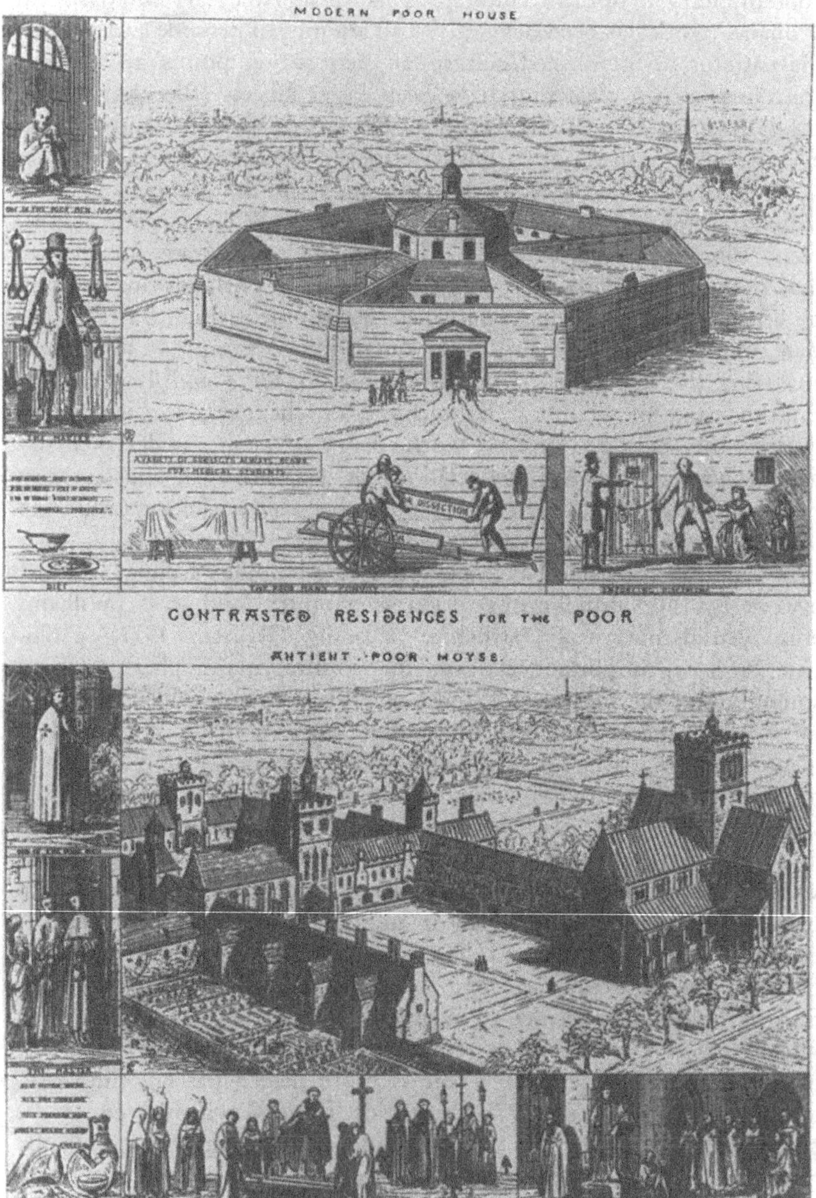

Figure 2.2 Contrasted residences for the poor

19th-century architect and critic, A. W. Pugin, further demonstrates the ability of cultural materialism to handle the complex interrelations of aesthetic, moral, and social issues. As did several of his contemporaries, including Carlyle and Ruskin, Pugin reacted to the advent of industrial society by looking backwards to the Middle Ages in search of a more organic society. The use of this medieval model to point out the perceived disadvantages of present-day industrialism is amply demonstrated in the polemical volume that Pugin entitled *Contrasts: or a parallel between the noble edifices of the Middle Ages and corresponding buildings of the present day, shewing the present decay of taste* (1836). Pugin's title accurately reflects his essentially comparative method, using a series of architectural contrasts between medieval and 19th-century designs to reveal the superior features of the former and the negative qualities of the latter. His judgements are not merely aesthetic, however, but include the moral and social implications embodied in architectural form. In one case, Pugin contrasts an idealized 'ancient poor house' with the stark modern-day equivalent, modelled on Jeremy Bentham's utilitarian-inspired Panopticon (Fig. 2.2). The 'noble edifice' of the Middle Ages resembles an Oxbridge college with its spacious quadrangles, gardens, and trees enclosed by ecclesiastical buildings which are in complete harmony with the surrounding landscape. In this monastic setting, it is clearly implied, the poor are treated humanely; they are given a reasonable diet and allowed to preserve their dignity. In the modern equivalent, the poorhouse resembles a jail, designed for efficient surveillance from a central point, rather than for the welfare of the poor. The master of the poorhouse is depicted with whip and manacles, keeping his charges in squalid and undignified conditions, with only a meagre diet and without the prospect of even a decent burial.

A similar contrast is provided in Pugin's comparison of a 'Catholic town in 1440' with 'The same town in 1840' (Fig. 2.3). The medieval town is dominated by the slender spires of a dozen churches, and by the solid walls of the abbey which give the town its aesthetic and functional coherence. In its 19th-century equivalent, many of the churches have been demolished; several others have been converted to more functionally designed dissenting chapels. The abbey has been replaced by an ironworks and the open space in the foreground is now dominated by another Benthamite Panopticon. The modern town of 1840 also features a new parsonage and pleasure-ground on the site of a former churchyard, a lunatic asylum, and a Socialist hall of science, as well as a town hall, and concert room. The organic charm of the medieval town has been replaced by a sprawling and haphazard collection of industrial buildings. In both cases, the architectural

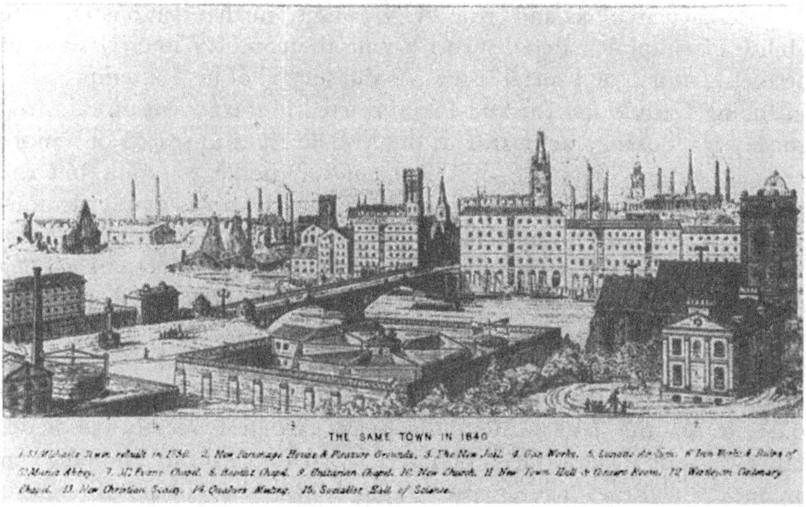

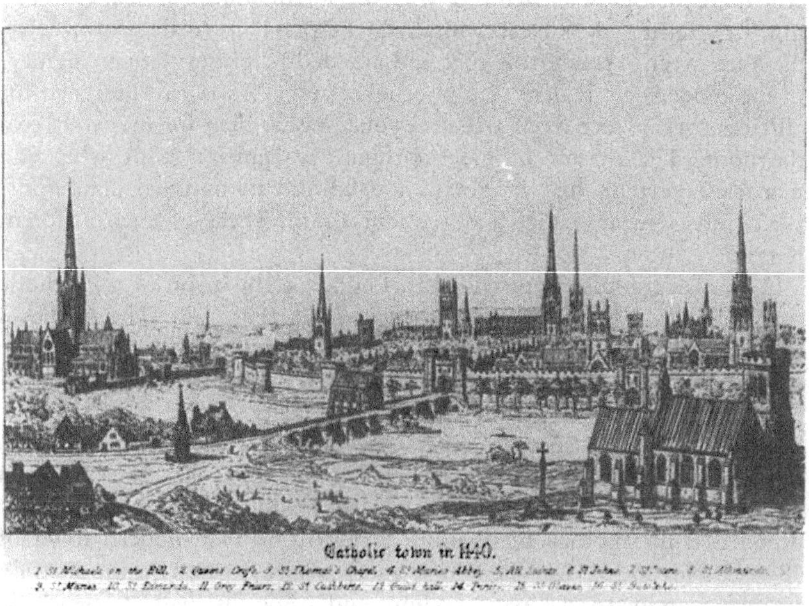

Figure 2.3 Contrasted towns of 1440 and 1840

contrast clearly implies a commentary on contemporary social values and moral standards. Culture and society could scarcely be more closely juxtaposed (Williams 1958, ch. 7).

A materialist cultural geography?

The previous examples indicate the potential for a materialist cultural geography, taking its inspiration from the work of Raymond Williams and other radical cultural theorists. Such a move has already begun among those geographers who have adopted a more critical conception of culture (cf. Cosgrove 1983, Thrift 1983). The final section of this chapter reviews some of the work that sets out to reconstruct cultural geography along broadly materialist lines.

In his essay on the origins of linear perspective and the evolution of landscape, Denis Cosgrove welcomes the revival of interest in landscape among humanistic geographers but proffers a number of caveats against the kind of landscape research that focuses exclusively on the subjective aspects of human experience, creativity and imagination (Cosgrove 1985a). Tracing the history of the landscape idea, Cosgrove offers a critical alternative to the extreme subjectivity of much landscape research, reinserting the landscape into contemporary political and ideological debates. Borrowing a phrase from John Berger (1972), Cosgrove shows how the concept of landscape developed as a bourgeois 'way of seeing' during the 15th and 16th centuries, rooted in the spirit of Renaissance humanism and in the exercise of power over land. He shows how linear perspective, in particular, employs the same geometry as merchant trading and accounting, navigation, land survey, mapping and artillery, and hence may be linked to the development of mercantile capitalism itself:

> The mathematics and geometry associated with perspective were directly relevant to the economic life of the Italian merchant cities of the Renaissance, to trading and capitalist finance, to agriculture and the land market, to navigation and warfare (ibid. p.50).

He goes on to show that the evolution of landscape painting and landscape gardening in Tudor, Stuart, and Georgian England can similarly be related to changing social relations on the land. The idea of a visually pleasing prospect, for example, coincided with the period when command over land was being established on new commercially-run estates by Tudor enclosers and the new landowners of measured monastic properties (ibid. p.55). The ideas of prospect and perspective can therefore be interpreted as a visual appropriation of space that corresponds to the material appropriation of land. By subjecting a traditional geographical concept such as landscape to a materialist analysis, Cosgrove radically alters the questions that can be asked of such an apparently familiar theme.

Stephen Daniels develops several related themes in his work on the landscape gardener, Humphry Repton (Daniels 1982), showing that landscaping was a highly contentious practice in late Georgian England involving debates that went far beyond the aesthetic qualities of landscape design. Daniels shows, for example, how Repton's designs for Sheringham Park in Norfolk confronted broad social issues concerning the power of the landed gentry as well as thorny moral questions concerning the rights and responsibilities of the landowning class in what was still an agrarian society. Repton's earlier commission at Armley, near Leeds, raised similar questions of landscape etiquette (Daniels 1981). In this case, Repton's disenchantment with the Romantic conception of the picturesque led him to undertake a series of extraordinary 'improvements' including the incorporation of a woollen mill, as well as a more conventional ruined abbey, in one of the house's principal vistas. The park's unconventional design, which Repton himself apparently found somewhat embarrassing in later years, suggests an artistic resolution of complex and contradictory attitudes towards industrialism that reflects the underlying social tensions of the period. Daniels concludes by arguing that greater attention should be paid to the material constitution of landscape images, rejecting the separation of landscape tastes from the business of production that has characterized more traditional readings of the cultural landscape.

Finally, Nigel Thrift (1983) returns to the theme of landscape and literature to explore the possibility of a materialist alternative in a field that has been dominated by humanistic geographers. He employs the concepts of 'hegemony' (from Gramsci) and 'structure of feeling' (from Raymond Williams) in an attempt to relate the representation of place in literature to wider cultural processes. Taking as examples the representation of the front in World War I and the significance of place in John Fowles' novel of English middle-class life, *Daniel Martin* (1977), Thrift explores the relationship between 'lived experience' and 'literary signification'. He emphasizes that the interpretation of culture in an aesthetic sense cannot be divorced from the production of culture in a material sense. Every literary representation of place is therefore an inherently political creation, just as every reading of a text offers the possibility of challenging received ideas about the politics of place.

Conclusion

Beginning with Tylor's problematic view of culture in 19th-century anthropology, this chapter has highlighted the shortcomings of 'culturalist' approaches to a range of issues. The weaknesses of culturalist approaches to 'urban culture' and to the 'culture of poverty' have been discussed and alternative approaches explored, focusing on the material process of cultural production. Insistence on a materialist approach to culture does not, however, involve a return to the 'material elements of culture' as emphasized by members of the Berkeley School (discussed in Ch. 1).[10] A broader conception of culture is required than one which limits its attention to physical artefacts and landscape features. Breaking out from traditional views of culture and landscape involves an analysis of the nature of ideology and its significance for social relations of production and reproduction. It implies a thoroughly politicized concept of culture and turns attention to areas of social life that have rarely been treated by geographers. These questions of cultural politics are taken up in the next chapter, which explores the notion of culture as ideology, and in the following chapters on popular culture, gender, sexuality, and race.

Notes

1 For example, there is no serious discussion of culture in *Models in geography* (Chorley & Haggett 1967).
2 As Althusser argues, 'a word or concept cannot be considered in isolation; it only exists in the theoretical or ideological framework in which it is used: its problematic' (1969, p.252).
3 For comparable British examples, see Hazel Flett's analysis of the way housing managers' discretionary powers produce unintentionally discriminatory effects (Flett 1979). Recent work by Henderson & Karn (1987) also reveals how formal rules and informal practices in housing allocation consistently reproduce inequalities between black and white tenants, despite well-intentioned but ill-founded notions of 'ethnicity' and 'preference'.
4 Roy Kerridge provides a comparable British example, describing Rastafarianism as 'a religion fit for wayward men', characterized by 'childish beliefs', and suited to 'loose-livers' who have 'several common law wives, often under the same roof'. Kerridge also makes the distinction between 'true Rastas' and the rest, condemning their addiction to marijuana, 'a drug of indolence that produces mental and spiritual deterioration in habitual users' (Kerridge 1983, pp.77-98).
5 Malcolm Cross refers to a similar situation in Britain as 'the manufacture of marginality' (Cross 1982).

6 Althusser defines historical materialism as 'the science of social formations' which are themselves defined as '"concrete complex wholes" comprising economic, political and ideological practices at a certain place and stage of development' (Althusser 1969, pp.250-1). Althusser's ideas have been the subject of hostile and penetrating criticism (e.g. Thompson 1978). Many Marxists have accepted the need to reformulate the distinction between base and superstructure outlined in classical Marxist theory even if they would not accept Althusser's own approach to the problem of defining 'structures of dominance' that are determining 'in the final instance'.

7 He describes these years movingly in *Politics and letters* (Williams 1979). Other biographical material is taken from the volume on Williams in the 'Writers of Wales' series (Ward 1981) and from the obituary by Terry Eagleton in *New Left Review* (1988).

8 Most biographical accounts have not dealt satisfactorily with these contradictions in Morris' life which demand a more sophisticated conception of cultural politics than they generally employ. For a notable exception, see Thompson (1955) who provides the kind of historical context necessary for an appraisal of the tensions between Morris' art and his politics.

9 Bourdieu himself defines *habitus* as 'systems of durable, transposable dispositions', 'principles of the generation and structuring of practices and representations' (Bourdieu 1977, p.72).

10 The kind of cultural materialism associated with Raymond Williams should not be confused with similarly named projects by cultural anthropologists such as Marvin Harris (1980). Despite a 27-page bibliography, Harris does not cite any of Williams' work. Indeed, Harris considers 'historical materialism' to be one of several alternatives to his own version of 'cultural materialism' which seeks to add notions of reproductive pressure and ecological variables to the material conditions specified by Marx and Engels.

Chapter three
Culture and ideology

In the previous two chapters the inadequacies of conventional geographical definitions of culture were outlined and the contours of an alternative, materialist approach were traced. This chapter continues the argument by examining the relationship between culture and ideology, seeking to establish a link between the world of ideas and beliefs and the world of material interests. For, as writers as diverse as Raymond Williams and Louis Althusser have argued, the theory of ideology cannot be confined to the realm of ideas and beliefs. According to Althusser (1969), ideology refers to the 'lived relation' between people and their world. It is a severely practical domain where ideas and beliefs have definite material consequences. For Williams, however, culture cannot simply be reduced to ideology, narrowly conceived, because it is part of a social and political order that is *materially produced.* Criticizing bourgeois conceptions of politics, Williams argues:

> What is most often suppressed is the direct material production of 'politics'. Yet any ruling class devotes a significant part of material production to establishing a political order. The social and political order which maintains a capitalist market, like the social and political struggles which created it, is necessarily a material production. From castles and palaces and churches to prisons and workhouses and schools; from weapons of war to a controlled press: any ruling class, in variable ways, though always materially, produces a social and political order. . . In failing to grasp the material character of the production of a social and political order, this specialized (and bourgeois) materialism failed also, but even more conspicuously, to understand *the material character of the production of a cultural order* (William 1977, p. 93; emphasis added).

In outlining a theory of ideology and tracing its relations with the concept of 'culture', care must always be taken to maintain the link between the material and the symbolic. Such a link is central to the work of the Centre for Contemporary Cultural Studies at the University of Birmingham where culture is defined as the medium through which people transform the mundane phenomena of the material world into a world of significant symbols to which they give meaning and attach value (Clarke *et al.* 1976, p.10). Unlike most geographical approaches to culture, this definition is not limited to the realm of material things. It also includes the less tangible world of consciousness and experience. It is through the medium of culture that people's raw experience is made sense of socially, by being transformed into a world of significant symbols.

Adopting a materialist approach, this chapter argues that the explanation for any ideology or cultural practice must be sought in specific historical and geographical circumstances. History is conceived of not as the simple passage of time, but as a dynamic process in which cultures are actively forged by real men and women. Similarly, geography is conceived of not as a featureless landscape on which events simply unfold, but as a series of spatial structures which provide a dynamic context for the processes and practices that give shape and form to culture. The materialism that informs this chapter is therefore both historical and geographical in seeking to explore the simultaneously social and spatial process by which people 'handle' the changing raw materials of their lives. It examines the extent to which place is a significant component in the production and reproduction of culture. Before the argument can proceed, however, appropriate theoretical language must be deployed concerning the nature of ideology and the concept of hegemony.

The concept of ideology

The origins of the concept of ideology can be traced back to the end of the 18th century. But it is with Marx that it can be said to have come of age (Larrain 1979, p.34). The concept was most explicitly developed in Marx's early works, particularly *The German ideology* (1846). These early works, which are among the most accessible of Marx's voluminous writings, were intended as a sustained attack on Hegelian idealism. Part I of *The German ideology* begins with a critique of Hegel's disciple Ludwig Feuerbach, couched in the general terms of an opposition between the philosophies of idealism and materialism. Marx refers contemptuously to the 'illusions' of German ideology, giving the phrase an entirely negative connotation. He argues that none of the philosophers whom he lumps together as 'idealists' had considered the connection between philosophy and material reality, or investigated the relationship between their style of criticism and their actual material surroundings. Marx defined his own position as the reverse of theirs:

> The premises from which we begin are not arbitrary ones, not dogmas, but real premises from which abstraction can only be made in the imagination. They are the real individuals, their activity and the material conditions under which they live (Marx & Engels 1846 (1970 edition) p.42)

Whereas the idealists descended 'from heaven to earth', from the realm of ideas to that of their material conditions, Marx argued the exact opposite:

> In direct contrast to German philosophy which descends from heaven to earth, here we ascend from earth to heaven. That is to say, we do not set out from what men (*sic*) say, imagine, conceive, nor from men as narrated, thought of, imagined, conceived, in order to arrive at men in the flesh. We set out from real, active men, and on the basis of their real life-process we demonstrate the development of the ideological reflexes and echoes of this life-process. . . Life is not determined by consciousness, but consciousness by life (ibid. p.47).

Marx criticizes Feuerbach for refusing to see that superficial appearances can only be explained through their connection with the material world. He proposes to explain these connections historically, in the sense that the material world can only be understood as the historical product of particular social practices. Marx's own unique contribution to the theory of ideology was, however, to make the connection between the ideological realm and the division of labour. Where contradictions arising from class conflict could not be resolved at the level of practice, Marx argued, they were resolved ideologically, at the level of consciousness. Ideology, then, has a specifically negative connotation for Marx implying not just 'false consciousness' (the concealment of people's real interests from themselves) but also the concealment of one group's interests from other people. Ideology is of practical significance, therefore, in concealing interests and negating social contradictions. It serves a crucial role in the reproduction of society, disguising the inevitability of class conflict and representing the interests of the ruling class as the interests of the whole of society. Marx makes this point most explicitly in a famous passage from *The German ideology* concerning the relationship between the ruling ideas and ruling class:

> The ideas of the ruling class are in every epoch the ruling ideas, i.e. the class which is the ruling *material* force of society, is at the same time its ruling *intellectual* force. The class which has the means of material production at its disposal, has control at the same time over the means of mental production, so that thereby, generally speaking, the ideas of those who lack the means of mental production are subject to it. The ruling ideas are nothing more than the ideal expression of the dominant material relationships, the dominant material relationship grasped as ideas (Marx & Engels 1846 (1970 edition) p.64).

Although Marx later came to revise this rather stark connection with a more elaborate discussion of the role of the state and civil society (Urry 1981), it provides a baseline from which to embark on studies of cultural production, the relationship between élite and popular culture, and the nature of subcultural resistance to the power of the 'ruling class'. Before undertaking such a discussion, however, it is important to consider some of the subsequent refinements to the concept of ideology that have been developed from Marx's early work.

According to Raymond Williams (1977 pp.55–71), 'ideology' has a number of quite distinct meanings even within contemporary Marxist thought. It may refer to any system of beliefs that are characteristic of a particular class or group (such as 'bourgeois ideology' or 'Protestant ideology'). By extension, it may refer negatively to a system of illusory beliefs, false ideas or false consciousness, contrasted, by implication at least, with the true knowledge provided by Marxist science. Finally, 'ideology' may refer to the general process of the production of meanings and ideas. Althusser, for example, subscribes to a version of this latter usage, arguing that ideology cannot be distinguished from science by its falsity (for ideologies can be quite coherent and logical), but by the fact that the 'practico-social' predominates in ideology over the theoretical (Althusser 1969, p.251).

Raymond Williams provides a detailed analysis of how ideologies work in practice by means of 'characteristic selectivities' (Williams 1981, p.27). In other words, ideologies operate by systematically promoting certain meanings in preference to others according to the discernible interests of a dominant social group. In this sense, an ideology can be defined as the way in which ideas come to represent certain interests or to conceal them in a more or less consistent way. For example, the statement that 'women are the fairer sex' is ideological in the sense that what is ostensibly a flattering comment about the nature of women actually represents an identifiable set of patriarchal interests that serve to perpetuate the subordination of women by men (see Ch. 5). It also illustrates the characteristic role of ideology in mediating the relationship between groups that are fundamentally unequal in terms of power.

Ideology frequently takes the form of 'commonsense' – ideas that are sufficiently 'taken for granted' as to be beyond the realm of rational debate. Many English people, for example, share an unquestioning belief in the stupidity of the Irish, the avarice of the Welsh, and the meanness of the Scots. Whether or not they really believe these stereotypes, they routinely reproduce them through day-to-day actions such as telling racist ('ethnic') jokes. In this sense, ideology is a form of 'unexamined discourse' (Gregory 1978, p.63) that goes well beyond the level of 'things said' as ideas become institutionalized in practice.

In seeking clarification of the diversity of meanings that have been applied to the concept of ideology, John Urry's restricted definition of ideology as the 'concealment of interests' is particularly useful (Urry 1981). According to Urry, statements are ideological insofar as they conceal the interests of a dominant group in one or more of the following ways:

(a) by *externalizing* social practices (e.g. by blaming a problem on forces 'beyond our control', such as when current unemployment levels in Britain are blamed on Commonwealth immigration, or on global economic recession, absolving the government from accepting its own responsibility);
(b) by *isolating* social practices (e.g. by speaking of civil unrest in British cities as a 'crisis of ethnic criminality that is not Britain's fault', as Peregrine Worsthorne wrote in the *Sunday Telegraph* (29 November, 1985), suggesting that the 'disorders' were solely the responsibility of one 'ethnic' group);
(c) by *conflating* social practices (e.g. by running together two or more ideas that are analytically separable, as when present-day inequalities between blacks and whites are explained purely in terms of past events such as slavery and empire, deflecting the blame for the effects of contemporary racism on to past injustices);
(d) by *obscuring the causes* of social practices (e.g. blaming the poor quality of inner-city housing on the residents of such areas rather than seeing the quality of housing as the result of economic discrimination, 'red-lining', and similar practices);
(e) by *obscuring the interrelations* between social practices (e.g. by denying the fact that racial stereotypes serve class interests: if the Asian workforce is regarded as inherently 'passive' then labour unrest can be blamed on the subversive activities of a few outside 'agitators'); or
(f) by obscuring *conflicts of interest* (e.g. by appealing to the 'common good' or to the idea that 'everyone agrees' with something when it

is clear that genuine consensus is very rare: communism is 'un-American'; the 'British people' are tolerant of foreigners but react adversely to being 'swamped' by immigrants).

Although most of these examples have been taken from the field of 'race relations', they could be replicated for any area of discourse involving an unequal distribution of power between two or more social groups, for these are the conditions in which ideology flourishes.

Having discussed various definitions of ideology it is important to be explicit about the usage adopted here. In describing the close relationship between culture and ideology, the negative implication of ideology introduced by Marx is retained. But it is still possible to use the term in a critical sense without implicitly contrasting 'ideology' with the 'truth' of an alternative (Marxist) science. Any statement of belief or any social practice can be regarded as 'ideological' insofar as it fails to make clear the interests that it represents. Ideologies offer decontextualized readings of social situations which are partial in both senses of that term (biased as well as incomplete). It follows that there is no single 'true' representation but many representations, each bearing its own ideological burden and each serving particular interests. Pierre Bourdieu suggests a similar understanding in referring to an ideology as 'an illusion, consistent with interest, but *a well-grounded illusion*' (Bourdieu 1984, p.74; emphasis added).

Such formulations may be criticized for their open-endedness and apparent ambiguity, But these are the very qualities with which contemporary cultural studies must be prepared to engage (cf. Levine 1985). Ideologies operate in diverse and subtle ways, but this need not imply that every interpretation is as valid as any other. In order to avoid an undisciplined relativism, a more critical theory of ideology is needed that distinguishes dominant ideologies from subordinate ones, recognizing that not all readings have the same power to persuade. As J.B. Thompson (1984 p.76) argues, in the case of several recent theorists, the concept of ideology has lost its critical edge through its separation from the *critique of domination*. To preserve the connection between ideology and power requires a discussion of the concept of hegemony, as advanced by the Italian Marxist, Antonio Gramsci.[1].

Hegemony and power

One of the most significant advances in Marxist thought stems from Gramsci's reworking of the concept of hegemony. In common usage, hegemony refers to a situation of uncontested political supremacy. In

Gramsci's work, however, it has a rather different meaning, referring to the power of a dominant class to persuade subordinate classes to accept its moral, political, and cultural values as the 'natural' order. In this sense, hegemony refers to the power of persuasion as opposed to the power of coercion through the use of physical force. Significantly, from the point of view of a ruling class, the exercise of hegemony is a much more efficient strategy than coercive control, involving the use of fewer resources and reducing the potential for open conflict by securing the acquiescence of the oppressed to their subordination.

The real innovation in Gramsci's work was the realization that, in capitalist societies, hegemony is never fully achieved – *it is always contested*. However powerful the élite become, their dominance will always be challenged by those in subordinate positions. Resistance may not always be active and open. Often it will be latent and largely symbolic. The ruling class may seek to limit the expression of resistance but, according to Gramsci, it will never be able to eradicate it entirely. Central to Gramsci's analysis, then, is a conception of the plurality of cultures. The dominant mores of any social élite may be aspired to by the middle classes but they are just as likely to be rejected by those lower down the social hierarchy as unobtainable, if not undesirable. In turn, members of the élite regularly condemn popular culture for its alleged vulgarity (see Ch, 4), seeking to establish their hegemony by suppressing anything they choose to define as cultural insubordination. Taking this argument further, some authors have inferred the existence of a hierarchical ranking of cultures, standing in opposition to one another in relations of domination and subordination along a scale of 'cultural power' (Clarke *et al.* 1976, p.11).

There are, however, problems with this formulation (not least the tendency to reify cultures in an unacceptably functionalist manner). For it is people (aligned as classes or within other social groups), not disembodied 'cultures', that oppose one another. And if it is difficult to rank social classes in terms of their differential access to material resources, or in terms of status and power, it is even more problematic to arrive at an agreed scale of 'cultural power'. One means of overcoming this difficulty is suggested by Frank Parkin's notion of *social closure*, a Weberian development of Marxist class theory (Parkin 1979). Rather than attempting to define social classes on *a priori* grounds (such as in terms of their differential access to the means of production), and then attempting to translate these abstract categories into actual social groupings (such as bourgeoisie and proletariat), Parkin prefers to define social classes in terms of the way in which they wield power. Significantly, this is the definition that Mark Billinge adopts in his application of the concept of hegemony to the historical geography of 18th- and 19th-century Britain (Billinge 1984), where hegemony is defined in terms of social power and culture is reinstated as an active channel in the struggle between capital and landed interests.

According to Parkin, dominant social groups are characterized by their ability to exercise power in a downwards direction, excluding less powerful groups from resources over which dominant groups exert control and to which they have privileged access. Parkin calls this process *exclusionary closure*, typified, for example, by the use of academic credentials to restrict access to certain occupations and professions where formal 'qualifications' are required. Subordinate social groups do not have this privilege and are forced to seek power in an upwards direction, attempting to make inroads into the resources controlled by more powerful groups. Parkin calls this *usurpationary closure*, typified by strikes and industrial action or by riots and rebellions.

Even within the sphere of conventional politics, subordinate groups have evolved a wide repertoire of strategies for resistance, negotiation, and struggle. This informal culture of the workplace includes attempts to exercise day-to-day control over the labour process, debates about the minimum wage for particular industries, as well as the 'down tools' strategy, the walk-out, the strike, the official dispute, and the factory occupation (Clarke *et al.* 1976, pp.41-2).[2] But there is also an almost inexhaustible number of ways in which subordinate groups can use cultural or symbolic strategies to resist subordination, and it is this field that the Centre for Contemporary Cultural Studies has really made its own. It should be immediately apparent that the exercise of exclusionary social closure can command the full resources of the state, the courts, and other institutions of law and order, while usurpationary strategies are much more likely to be considered illegitimate, if not downright illegal. This is one reason why resistance often takes a symbolic rather than a directly instrumental form.

There are further problems in defining what constitutes a 'dominant class' in contemporary capitalist societies. In Britain, for example, there are reasons for regarding the aristocracy and the bourgeoisie as representing different fractions of the dominant class, united only insofar as they are opposed to the emancipation of the working class. Similarly, the working class can only rarely be defined as a united body with a homogeneous and neatly defined set of common interests. It is notoriously subdivided along lines of race and gender, with white working-class men seeking to exercise power over women and blacks by means of exclusionary practices that restrict access to certain jobs.[3] In Parkin's terms, these divisions within the larger 'excluded' group can be conceptualized in terms of 'dual closure' (see Fig. 3.1) where those who are subject to exclusionary closure in turn seek to exclude others who are less powerful than themselves.

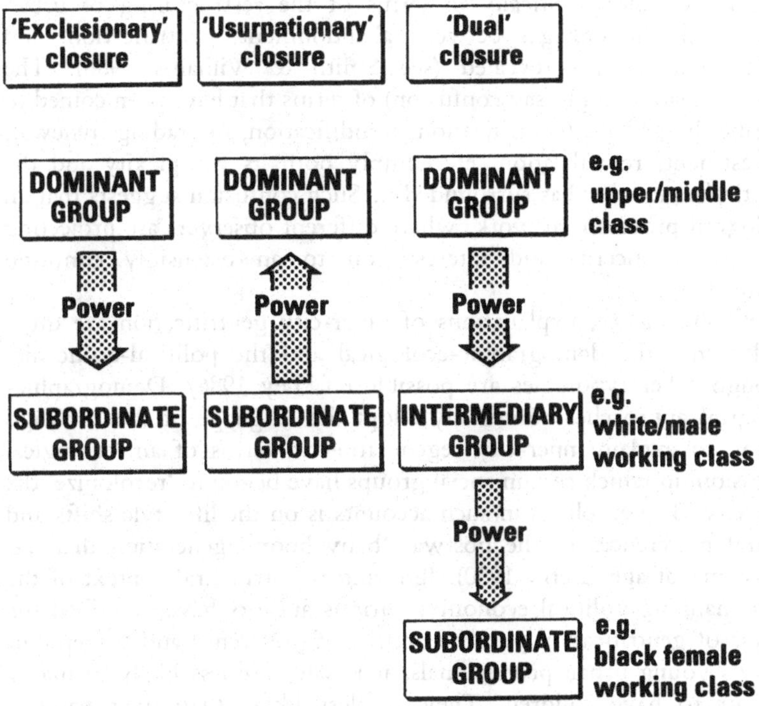

Figure 3.1 The theory of social closure

The important point, however, in the present context is not to arrive at some agreed definition of what constitutes the 'dominant social group' but to discover how particular groups achieve positions of relative power, how they seek to maintain power by successfully dominating subordinate groups, and how those groups themselves contest their subordination. A series of examples shows the subtle ways in which ideology operates in contemporary capitalist societies, suggesting that cultural strategies may be as important as economic and political ones in maintaining the *status quo*.

Culture and capital in urban change

The first example concerns the ideological character of much of the recent literature on gentrification, variously described as an 'urban renaissance', a return of the middle class to the city, or a triumphant display of 'urban pioneering'. Only recently has the process been reinterpreted more critically in terms of the restructuring of urban space, with the linkages between abandonment, gentrification, and displacement clearly revealed (see Smith & Williams 1986). The profusion (some might say confusion) of terms that have been coined to describe the process (gentrification, trendification, upgrading, renewal, reinvestment, revitalization, etc.) imply both its complexity and the analytical chaos that has surrounded it. Such confusion suggests that an ideological process is at work, where different observers are projecting their own concerns and interests on to an ostensibly common phenomenon.

For convenience, explanations of inner-city gentrification are often divided into the demographic-ecological and the political-economic, although other typologies are possible (cf. Ley 1986). Demographic-ecological approaches commonly adopt the language of the 'Chicago School' to explain inner-city regeneration in terms of an 'ecological succession' in which certain social groups have begun to 'recolonize' the inner city. The emphasis in such accounts is on the life-style shifts and cultural preferences of the postwar 'baby boom' generation that has now come of age (Berry 1980). Ignoring the structural context of the city's changing political economy, various authors have described the process of gentrification purely in terms of preference and perception. Today's young urban professionals, it is said, are less likely to marry early or to have children. They are less likely than their parents' generation to reject the ageing, congested, crime-ridden, and polluted cities in search of a dream home in the suburbs and are coming 'back to the city' in search of a new urban life-style. The fact that both the earlier suburbanization trend and the current trend towards gentrifying urban neighbourhoods have been heavily subsidized by the state goes virtually unremarked, except by more radical analysts (e.g. Smith 1979a, 1979b, 1982). For preferences cannot be analyzed apart from opportunity; demand cannot be assessed in isolation from supply. But this is exactly what the 'back to the city' literature ignores, particularly in its more popular, journalistic form.

To cite some examples, a few years ago the *New York Times Magazine* ran a cover story entitled 'Rediscovering the city: the new élite sparks an urban renaissance' (14 January 1979); other popular writers describe gentrification using such romantic titles as *Pioneering in the urban wilderness* (Stratton 1977); and bona fide social scientists argue that: 'Basically gentrification stems from the strong desire of a significant number of upper-middle-income households to remain physically close to the mix of economic, social, and cultural opportunities which are uniquely found in the central business districts of larger and older cities in the U.S.' (Tobier 1979, p.14). The ideological effect in each case is to obscure the structural roots of gentrification and to conceal the very real conflicts that the process entails. 'Gentrification' is, after all, a term that focuses on the changing class structure of the inner city.

Some ecological analyses of gentrification take a more sophisticated form, searching for the roots of contemporary urban change in wider cultural movements. An example is provided by Charles Simpson's analysis of the growth of artists' living quarters in the studio-lofts of New York's SoHo district, South of Houston Street in lower Manhattan (Simpson 1979). Despite his sensitivity to the 'sacred' associations of artistic production which generated a sympathetic political attitude towards the artist-tenants in the early days of SoHo's transformation, Simpson accounts for the subsequent creation of a luxury real-estate market for residential lofts as arising from the needs of the 'culturally sophisticated middle class' whom he describes as 'returning to urban residence in search of a varied and individuated life-style unavailable in the suburbs' (ibid. p.4). In fact, most recent evidence suggests that the gentrification trend involves an established urban population rather than a returning suburban one (cf. Jackson 1985). Similarly, the emphasis on cultural sophistication and new urban life-styles, is readily countered by other analyses that emphasize the intersection of 'culture and capital in urban change' (cf. Zukin 1988b). Sharon Zukin explains the development of the real estate market for luxury lofts in terms of the opportunities for redevelopment created by New York City's declining manufacturing base, aided by the compliance of the city's patrician interest in historic preservation and the arts, the intervention of the state with respect to zoning and building codes, and the extension of key fiscal incentives to private redevelopers. In such circumstances, there is considerable support for those who feel that American gentrification has been 'actively planned and publicly funded' (Smith 1979b).

Not surprisingly, government agencies and public officials deny their active role in promoting gentrification. From the perspective of New York's Department of City Planning, for example, 'private reinvestment' in Brooklyn's Park Slope district and in Manhattan's Upper West Side has been 'a positive influence' on both neighbourhoods (New York Department of City Planning 1984, p.vii). Characteristically, the extent of residential displacement is denied, downplayed, or regarded as unquantifiable. A number of ideological ploys can be adopted to maintain this position as the following examples (from Laska & Spain 1980) illustrate.

The most common ideological device is to contrast the negative effects of housing abandonment and physical decay in the inner city with the alleged benefits of gentrification. This strategy is apparent in the language of gentrification which implies that the advantages of regeneration, reinvestment, and renovation are now taking over from the disadvantages of degeneration, disinvestment, and decay that apparently preceded them. What goes unquestioned is how the benefits and adverse effects of gentrification are distributed across the population. Who gains and who loses from contemporary urban change? What aspects of the 'inner-city problem' are 'solved' by gentrification, and what new problems are created? By ignoring such fundamental questions, these analysts imply that gentrification is to be welcomed irrespective of its social costs:

> The back-to-the-city trend has been seen as an unexpected hint of a reversal of decline which had been depicted by some as hopeless and irreversible. Because it offers just such a hope it is likely that most Americans believe the renovation trend should be encouraged at every opportunity, *regardless of the problems which accompany it* (Laska & Spain 1980, p.xvi; emphasis added).

The reference to 'most Americans' even implies that it is 'unAmerican' to oppose gentrification. The tenor of this kind of specious argument is reproduced in more invidious form in other contributions, notably that by Howard Sumka, an official of the federal Department of Housing and Urban Development (HUD). Citing evidence of continued urban decline, he argues that gentrification is in the public interest: abandonment is the real problem and any form of reinvestment must be encouraged. Cities cannot survive as 'reservations for the poor'. Sumka's argument continues, with the reassurance that people have always been displaced in the 'natural' course of neighbourhood change and that, in the absence of firm evidence to the contrary, gentrification should not be regarded as having exceptionally harsh consequences. Indeed, gentrification appears to be a panacea for all the city's ills if, as Sumka argues, urban reinvestment 'holds out the prospect of regenerating the nation's central cities, conserving existing capital investment, aiding energy conservation, promoting the conservation of suburban and urban land, and restoring local fiscal balance' (Sumka 1980, p.278). Adding insult to injury, Sumka concludes that those who are displaced may be able to improve their housing conditions by moving to more desirable neighbourhoods! Reinvestment is a 'fragile process' with which the federal government should not tamper. He paints a touching picture of the attenuated powers of the state and concludes that, in neighbourhoods where the private sector is the moving force, 'there is little the federal government can do to slow the process' (ibid. p.283). In a few slick phrases, Sumka has denied the social costs of gentrification, implied that the process is the result of pure market forces, and denied the regulative role of the state in mediating the consequences or directing the future of urban neighbourhood change. Recognizing the ideological dimension of the analysis robs it of much of its power.

As this example shows, the critique of ideology provides a means of challenging received wisdom by exposing the interests that it serves. However dominant an ideology may seem, alternatives can always be sought by those who wish to resist its claims to hegemonic status. The next section attempts to interpret these challenges to the dominant ideology in terms of the concept of *resistance*, also developed from the writings of Antonio Gramsci.

Rituals of resistance

Despite the fact that many strategies of resistance are defined officially as illegitimate, 'usurpationary' in Parkin's terms, resistance does not inevitably take the form of active struggle. Often, indeed, the meaning of resistance is latent and appears purely symbolic in form. The Centre for Contemporary Cultural Studies has catalogued a range of working-class subcultures in postwar Britain, from Skinheads and Teds, to Rastas and Rudies, all of which they describe in terms of *rituals of resistance* (Hall & Jefferson 1976). By 'rituals', a coherent set of actions is implied the meaning and purpose of which are symbolic rather than purely practical, and which are routinized in the sense that they can be practised almost unconsciously. These rituals include various styles of dress and patterns of verbal and non-verbal behaviour the adoption of which implies an attitude of resistance to those in power.

Whether such 'rituals of resistance' can be regarded as truly *strategic* is a matter of some debate. A 'strategy' implies at least some degree of conscious deliberation rather than an idiom that is routinely and unreflectively employed.[4] For this reason, 'style' may be a better term than 'strategy'. Style is a secret language (Chambers 1986), and can be a form of cultural insubordination that expresses an attitude of defiance and disrespect to those in authority. Stuart Hall describes how one such style developed among young Jamaicans arriving in Britain in the 1950s, an image that was captured by the photo-magazine *Picture Post* (Fig. 3.2):

> Jamaicans travelled – as they went to Church, or to visit their relatives – in their 'Sunday best'. . . The clothes are those of someone determined to make a mark, make an impression on where they are going. Their formality is a sign of self-respect. . . These folk mean to survive. The angle of the hat is universally

Figure 3.2 The evocation of style (*Picture Post, 1956*)

jaunty: cocky. Already there is *style* (Hall 1984, p.4; emphasis in original).

As this passage suggests, rituals must be read in context and as an ensemble if their meaning is to be clearly understood. There is nothing intrinsically threatening about the wearing of boots rather than shoes, or of a shaved head as opposed to long hair. They only come to signify aggression and rebelliousness when they are given symbolic meaning – when they represent a style of dress and a way of life that differs from the mainstream. Indeed, many of the symbols of postwar working-class subcultures have been appropriated from the world of middle-class conformity and relative affluence. But once materially *appropriated* and symbolically *transformed* as part of a particular subcultural style, they take on new, sometimes threatening meanings.

Dick Hebdige has produced a number of studies that trace the process by which the meaning of objects is transformed through the appropriation of particular commodities that have previously enjoyed a more 'secular' use (Hebdige 1976, 1979). The symbolic transformation of the motor scooter by the Mods is a classic example (Hebdige 1983, 1988). Originally designed to be ridden by decorous Italian women for whom motorbikes were considered inappropriately masculine, the scooter became a central symbol of a subcultural style, representing a particular form of youthful defiance that was structurally opposed to the 'hard' image adopted by the motorcycle-riding Rockers (see Fig. 3.3).

Resistance may also be discerned in less tangible ways, such as in the variety of linguistic forms that different groups adopt to mark out a space for themselves, setting up a boundary with other social worlds (see Ch. 7). The city comprises a mosaic of social areas, each of which may develop its own distinctive pattern of speech, its own argot or subcultural vocabulary. In London in the early 1980s, for example, the 'Sloane Ranger' emerged as a temporary cultural icon, resulting from the conjunction of particular patterns of conspicuous consumption, language, and dress (nouvelle cuisine, Porsche cars, upper-class accents, Barbour jackets, striped shirts, and pearl necklaces) with particular neighbourhoods (Sloane Square, Kensington and Chelsea). They have since been replaced by a more generalized 'yuppie' image imported from the United States and with a more diffuse geographical base, though recognizable from Wall Street to London Docklands.

Generally, of course, subcultural forms are associated with subordinate rather than with dominant groups. Within the Chicago School of urban sociology, for example, Robert Park divided the city into a series of what he described as 'natural areas' each of which was characterized by 'its own peculiar traditions, customs, conventions, standards of decency and propriety, and, if not a language of its own, at least a

MODS	**ROCKERS**
Scooter	**Motorcycle**
Italianness	**Britishness**
Femininity	**Masculinity**
Surface	**Depth**
Consumption	**Production**
Leisure	**Work**
Service	**Heavy industry**
Art	**Science**
Soft	**Hard**

Figure 3.3 Symbolic oppositions of Mods and Rockers

universe of discourse, in which words and acts have a meaning which is appreciably different for each local community' (Park 1952, p.201). Park's students went on to describe these local areas, each with its own distinctive idiom, in great ethnographic detail. Classic examples include Nels Anderson's *The hobo* (1923), in which subtle distinctions were drawn between the 'main-stem' and the 'jungles', and between 'dead line men', hobos, tramps, and bums; and Paul Cressey's *The taxi-dance hall* (1932), where young girls pursued a moral career that degenerated rapidly from 'monkey hops' and 'stag dances' to a life of semi-prostitution in the down-market 'black-and-tans'.

As the previous examples suggest, resistance often takes a specifically territorial form. Thus it is possible to interpret the social and spatial structure of working-class neighbourhoods and of black ghettos, for example, in terms of resistance to the oppressions of race and class. The characteristic social forms of working-class neighbourhoods, parodied in *Coronation Street* and *East Enders*, with their corner shops, pubs, and other symbols of 'community', provide a relatively autonomous social space in which to seek respite from the all too pervasive influence of the factory or workshop. Working-class communities have always provided a symbolic space in which the tensions between work and leisure are played out. Thus, in the 19th century, it was not uncommon for pioneer industrialists to try to control the leisure time of their employees. Several of the earliest experiments in town planning bear the marks of such 'enlightened self-interest': New Lanark boasts an Institute for the Formation of Character; Port Sunlight has an art gallery, allotment gardens, and recreation grounds; and Saltaire has a chapel, an institute, and a library (but no pubs). Though in the factory, mill, or mine, the capitalist was indisputably boss, at home working-class families were able to maintain at least some degree of autonomy over their lives (Clarke & Critcher 1985). Maintaining even this much autonomy involved tortuous relations of class and gender, the job of 'holding the family together' often falling to the mother who was responsible for managing the household budget and who could assert considerable authority within the home (Hoggart 1957, Young & Willmott 1962).

The pattern was even more complicated in North America during the early part of this century with black ghettos beginning to emerge as partially separate worlds, defined principally in terms of race and class. Excluded from the rewards of status and power that derive from full participation in mainstream (white) society, black people created their own neighbourhood institutions to service their needs. As with the British working class, however, the degree of autonomy of such communities can easily be exaggerated. Frequently the main institutions – newspapers, stores, theatres, and other commercial establishments – remained under the control of white owners and managers. Any attempt by blacks to move beyond the confines of the ghetto provoked immediate, fierce reprisals, as Osofsky (1966) and Spear (1967) have documented in the case of New York and Chicago respectively.

These territorial battles can be interpreted from both the 'insider's' and the 'outsider's' perspective. From the 'inside', they represented an opportunity to challenge the economic hegemony of white society. A black bourgeoisie emerged in the ghettos of the north as blacks moved into the professions or became landlords and shopkeepers in their own right (though largely confined to providing homes and services for other blacks). But, by gaining an economic foothold in this restricted social space, serving a limited (and generally impoverished) clientele, black people virtually guaranteed their continued exclusion from the majority white society. From the latter's perspective, of course, the reverse was true: ghettos and working-class communities represented a very effective means of 'social control'. The winning of 'ghetto space' may represent something of a Pyrrhic victory therefore, unless it provides a platform from which to launch an assault on the wider society.

In Britain, Paul Willis' study of 'how working class kids get working class jobs' (Willis 1977) reaches similarly depressing conclusions. Tracing the passage of a group of young men from school to work, he shows how their unequivocal rejection of middle-class values represents a symbolic opposition to their structural subordination. Their defiant attitude to school leads them directly into working-class jobs and perpetuates their exclusion from middle-class occupations. The very patterns of behaviour that Willis interprets as resistance lead these particular kids to collude in their own oppression.

What alternatives are there to these pessimistic conclusions? Are all forms of resistance doomed to remain 'merely symbolic' in the sense that they do not have any material consequences? The effectiveness of symbolic resistance can often be judged by the extent to which it is perceived as a threat by those in authority. Even such intangible cultural forms as music are capable of expressing rebellion with such force that they provoke opposition from the self-appointed guardians of public morality and 'good taste'. Thus, both John Street (1986) and Paul Gilroy (1987) have written about the cultural politics of rock music, focusing on its ability to challenge conventional bourgeois values and on attempts by the state to regulate its more rebellious aspects.

Though few instances of counter-hegemonic resistance are purely symbolic in form, a number of writers have described such strategies as 'imaginary' or 'magical' (Hall & Jefferson 1976, Hebdige 1979). Where answers to people's practical problems of finding a job and making a living cannot be found on a material (economic) level, and where political action is not contemplated, resistance understandably takes a 'cultural' (symbolic) form: 'There is no "subcultural solution" to working-class youth unemployment, educational disadvantage, compulsory miseducation, dead-end jobs, the routinization and specialisation of labour, low pay and the loss of skills' (Clarke *et al.* 1976, p.47). It is no surprise, then, that people cannot 'solve' these problems in a practical sense. Instead, solutions emerge that express resistance in an 'imaginary' (symbolic) way, their political content being expressed in a cultural form.

Notwithstanding the 'imaginary' or 'magical' qualities of symbolic protest, it is also possible to analyze resistance in very practical terms via the actual symbols employed, most of which assume an immediate, tangible form. Take, for example, Dick Hebdige's analysis of the 'revolting' style of punk rock, with its insignia of safety-pins, chains, plastic bin-liners, and dyed hair, reflecting a voluntary identification with the position of social outcast and a wilful desecration of the socially approved values of middle-class style (Hebdige 1979, pp.106-12). As with Teddy Boys, Skinheads, and Rastas, the very symbols of resistance provide evidence of their social meaning once they are subjected to an appropriate reading.

This kind of symbolic analysis has been both challenged and defended in recent years. In the introduction to the new edition of *Folk devils and moral panics*, Stanley Cohen (1987) criticizes some of the more extravagant theories of subcultural style.[5] If subcultures are to be explained in class terms, for example, then why do similar class locations give rise to such a variety of responses and modes of accommodation? Why is so much subcultural analysis confined to such a narrow range of spectacular, masculine, working-class behaviour?

And why are the analysts of subcultural style prepared to make such allowances for their subjects' racist or sexist attitudes when they are so quick to condemn similar attitudes and behaviour in bourgeois culture? While Cohen applauds the recent tendency in cultural studies to reassert the historical and the political, he deplores the over-theorization of subcultural symbolism: 'the whole assembly of cultural artefacts, down to the punks' last safety pin, have been scrutinized, taken apart, contextualized and re-contextualized', as 'the conceptual tools of Marxism, structuralism and semiotics, a Left-Bank pantheon of Genet, Lévi-Strauss, Barthes and Althusser have all been wheeled in to aid this hunt for the hidden code' (ibid. p.ix).

It is no surprise, then, that one of the Left Bank's most eminent intellectuals, Pierre Bourdieu, has provided some of the best theoretical ammunition with which to defend the decoding of subcultural style. Speaking of the 'cultural competence' necessary to participate in any subculture, Bourdieu argues that: 'A beholder who lacks the specific code feels lost in a chaos of sounds and rhythms, colours and lines, without rhyme or reason' (Bourdieu 1984, p.2). He suggests that cultural competence is a 'cognitive acquirement'. As Bourdieu's own work confirms, however, understanding particular cultural codes in order to decode them requires detailed ethnography rather than ungrounded theorization. It is precisely this kind of evidence that is so conspicuously absent from discussions of football 'hooliganism' (the next example in this chapter) which has allowed all kinds of people to project their preferred reading on to a phenomenon for which there is little agreed empirical evidence.[6]

Folk devils and moral panics

From a materialist perspective, the success of any decoding depends on the analyst's ability to demonstrate how the outward manifestations of a cultural style are related to the political and social context in which they have emerged. Take, for example, the subcultural style of football 'hooliganism', which involves a range of practices including fanatical support of a particular team, ritualized chanting on the terraces, routine verbal abuse of rival fans, players, and officials, and occasional physical violence. (Other behaviour, such as pitch invasions and 'taking ends', seems to have declined in popularity, not least because of improved surveillance facilities and crowd control at football grounds.) In general, the press have treated football 'hooliganism' as an irrational phenomenon, an inexplicable aberration from society's unspoken moral codes. 'Hooliganism' is reported as 'soccer madness' and fans who take their enthusiasm to excess are 'football mad' (Taylor 1971). Occasionally, journalists attempt more 'scientific' explanations, referring to the amount of alcohol consumed before the match or to the psychology of crowds (invoking the notion of 'mass hysteria'). Invidious distinctions are also often drawn between the 'genuine supporter' (or 'real fan') and the rowdy minority who are blamed for all the 'trouble'.

Other types of explanation are, however, available which do not reflect the rhetorical excesses and crusading spirit of the popular press and which are more in keeping with contemporary theories of culture and ideology. Such explanations begin by identifying the historical conditions and structural context in which 'football hooliganism' emerged, coupled with an analysis of the specific features of 'football culture' which provide the raw materials for the subcultural style called 'hooliganism'. These explanations focus on why 'hooliganism' has become a problem here and now, and why it has not emerged on such a scale in other countries or in association with other sports, such as cricket or rugby league.

Identification of the appropriate structural context must be more rigorous than mere allusions to rising unemployment, the decline of manufacturing employment, and other 'inner-city' problems, important though these factors no doubt are. Social conditions in specific cities would need to be analyzed, probing the structural roots of traditional rivalries between different clubs. A lead in this direction has been taken by Bruce Murray's analysis of the interweaving of sectarianism, sport, and society in the traditional Glasgow rivalry between (Protestant) Rangers and (Catholic) Celtic (Murray 1984), an analysis that could be repeated for other clubs such as Liverpool and Everton.

A thorough understanding of 'soccer madness' would need to investigate why football matches have become popular recruiting grounds for the National Front and other extreme right-wing political parties such as the British Movement. It would need to differentiate between 'hooliganism' and other subcultural forms that emphasize physical violence as a definitive aspect of youthful masculinity. But, above all, it would need to be historical, even if not necessarily sharing Walvin's simple equation between the decline of Britain and the demise of the 'People's Game' (Walvin 1975, 1986). Only from an historical perspective could one begin to criticize the common-sense belief that football fans used to be less violent than they are today. An historical analysis would also allow one to distinguish some of the crucial relationships between sport, power, and ideology (Hargreaves 1986) and to make a more considered judgement about some of the supposed parallels between the media's recurrent obsession with 'hooliganism' and other 'moral panics' about Mods and Rockers in the 1960s (S. Cohen 1972) or about 'mugging' in the 1970s (Hall *et al.* 1978).

At least some of the blame for the current 'moral panic' about football 'hooliganism' can be attributed to qualitative changes internal to the sport itself. This view has been expressed in Ian Taylor's 'speculative sociology' of the origins of soccer violence (Taylor 1971) where he argues that football was, until recently, a traditional working-class sport that was characterized by a kind of 'participatory democracy' involving players, directors, managers, and supporters. Players came from the same working-class background as supporters and did not move significantly out of that class; success brought fame rather than fortune. Supporters rarely expressed antagonism towards the club's management which shared the fans' own intense loyalty to the local team. In recent years, however, this culture has been eroded by the professionalization and internationalization of the sport. Football has become a 'spectacle' with celebrities, media publicity, and an undisguised business orientation. Players and managers no longer share with supporters the same loyalty to the club. They are, instead, clearly motivated by financial gain and can be transferred to a rival team provided that the terms are right.

Sport, no less than other cultural forms, is subject to specific forces of production. In the case of football, for example, the interest of television companies in covering live matches is frequently said to have reduced attendance at matches, forcing clubs to increase the cost of admission. Clubs also sought to bolster their incomes by broadening their appeal away from their traditional working-class base. Additional seating and catering facilities were provided, together with 'hospitality boxes', further alienating the working-class supporter on the terraces. Changes in the social structure of the game were mirrored in the geography of the ground: stands replaced terraces, eating and drinking facilities were added, together with floodlights and all the paraphernalia of supporters' clubs. Some clubs even demolished terracing to accommodate new shopping facilities in an effort to supplement their dwindling income. As a result of these changes, Taylor concludes, a working-class 'rump' was formed which contested the embourgeoisement of the game by any means at their disposal, including violence. Their intervention was occasionally instrumental, such as invasions of the formerly 'sacred turf' to contest an adverse decision. More often, their intervention was 'magical' in the sense that it was not directed against any of the public representatives of the game, over which the supporters felt themselves to have lost control.

One does not have to agree fully with Taylor's conclusion that hooliganism is 'a "democratic" response to the loss of control exercised by a football subculture over its public representatives' (ibid. p.372) to sympathize with his general approach. Indeed, Taylor has himself revised his earlier argument, suggesting that greater emphasis should be placed on the broader relations between class and state: 'the key to the current decomposition of working-class spectator sport lies in the decomposition of the working class itself' (Taylor 1982, p.181). From this perspective, the 'respectable fears' of government officials and middle-class spectators towards football violence are more easily understood (cf. Pearson 1983). The Thatcher government's authoritarian reaction to recent outbreaks of soccer violence adds further weight to this interpretation. Additional policing at soccer matches, the banning of alcohol from football grounds, increased recourse to the courts and to international authorities, internal surveillance at grounds, and calls for visiting fans to carry identification cards all suggest that wider questions are being addressed through soccer violence and that 'hooliganism' cannot usefully be regarded as a pathological outbreak of inexplicable violence.

Football 'hooliganism' involves the construction of a typical 'folk devil' by the guardians of middle-class respectability who have reacted to a perceived increase in soccer violence by creating a classic 'moral panic' (S. Cohen 1972). It remains to be seen whether there is a specifically territorial basis to this kind of ideological crisis, through which a geography of resistance can be defined.

Territorial struggles

Responding to political and economic forces beyond their control, people frequently transfer the blame to a more readily identifiable local target. A classic instance is provided by Sean Damer's study of a multi-storey council estate in the Govan district of Glasgow (Damer 1974). The area, known locally as 'Wine Alley', was highly stigmatized by residents in more respectable neighbouring estates. Its 'dreadful' reputation arose in a context of extreme competition for the scarce resource of council housing on Clydeside at the height of the Depression. Based on the ignorance and suspicion that flourish in the absence of direct first-hand knowledge, the neighbourhood's reputation served as a convenient myth which helped to bolster Govan's waning moral community by treating Wine Alley as a stereotyped 'out-group'. According to Damer:

> It was a period of . . . 'boundary crisis', a period when Govan was ambivalent about its moral order, and, consequently, the Wine Alley people served as a handy demon against whom the community could be re-mobilised by its moral entrepreneurs. The scape-goating of the residents of the new estate played the function

for Govanites of re-asserting the boundaries of their community (Damer 1974, p.243).

A similar displacement process can be seen at work in much of the literature on race and crime, where fears about neighbourhood deterioration and increased crime rates are often expressed in terms of people's fears of 'ethnic' or 'racial succession' (Taub *et al.* 1984). Susan Smith provides the most plausible explanation of how race becomes the medium through which such fears are publicly articulated. On the basis of her research in the Handsworth district of Birmingham, she argues that ethnicity and race form the most overt means of handling routine interactions in an uncertain environment because they provide the most *visible* symbolic cues (Smith 1984a). In such cases, a process is set in motion whereby a territorial issue (such as neighbourhood crime) is encoded in racial terms. A stereotype is created that functions in a self-fulfilling way, reinforcing people's fears about the dangers of neighbourhood crime.

A third example concerns the development of London's Notting Hill Carnival, described by the anthropologist Abner Cohen as a 'contested cultural performance' (Cohen 1982). He shows how Carnival has taken a variety of forms, from an 'English Fayre' to a 'polyethnic Carnival', dominated first by Trinidadians and later by Jamaicans, expressing a variety of conflicts: between police and people, black and white, young and old, Jamaican and Trinidadian. Carnival, as described by Cohen, is an inherently political event which draws on a dynamic cultural repertoire: 'The cultural is structured by the political, though is not determined by it' (Cohen 1980, p.79). But a key feature of Carnival is its 'unruly' quality as public performance. As a form of unregulated street life, Carnival has defied the repeated efforts of the authorities to contain it within a sports stadium or similar arena (Jackson 1988a). Carnival expresses interweaving ideologies of race, class, and gender, constructed and contested through the symbolic language of music, dance, and public performance (cf. Owusu & Ross 1988). But it is the territorial dimension of Carnival that gives it such significance as a 'ritual of rebellion'.

With few exceptions, however, geographers have done little research on rituals of resistance, rarely moving outside the parameters of an out-dated sociology of 'deviance' (Becker 1963). The exceptions include some innovative work on urban graffiti and gang behaviour, under-taken by David Ley and others (Ley 1974, 1983, Ley & Cybriwsky 1974). According to these authors, gang membership offers a disreputable and extra-legal alternative route to the acquisition of status, denied them by more respectable and legal means:

> The gang serves a number of functions to black adolescents. It provides a strong peer group, identity, discipline, adventure, a measure of security, recognition, prestige. . . One of the greatest problems in turning gang members 'conservative' is to open new channels for winning prestige and status (Ley 1974, p. 128).

Ley proceeds to analyze a particularly brazen act of gang violence, involving the shooting of a rival gang member, in precisely these terms. The incident began, he argues, with teenagers denied access to legitimate outlets for recognition in mainstream society. Forced to invent alternative sources within their own life space, their solution was the enacted status of gang membership with, in this case, tragic consequences. The argument invokes a number of geographical principles to explain the incidence of violent gang behaviour within the inner city. The majority of incidents involving homicide, reported stabbing, shooting, or gang fights in the Philadelphia neighbourhood which Ley studied occur over very short distances between neighbouring gangs' home turfs (Fig. 3.4). In situations of this kind, proximity breeds aggression and increases the probability of violent conflict. Similarly, the spatial incidence of urban graffiti which Ley and Cybriwsky investigated in another Philadelphia neighbourhood can be explained in terms of tensions within the city's tight housing market where residential change is proceeding rapidly. The distribution of graffiti is concentrated at the centre of gang turfs, internally promoting the status of the gang, and at the edges of gang turfs, where space is actively contested and boundaries are constantly under threat (Fig. 3.5).

These are insightful analyses of urban behaviour and its spatial contours. Yet, sociologically, they contain some dubious assumptions. The implication that black and working-class subcultures are thwarted attempts to emulate middle-class values would not now go unchallenged. Subcultures possess a greater degree of autonomy than this suggests. Rather than seeing black or working-class cultures as 'deviant' forms of white, middle-class culture, a radical alternative would probe the structures of inequality that generate and legitimize these patterns of behaviour. For it is the structural dimensions of gender, race, and class that produce the social space between groups within which subcultural styles are elaborated.

An impressive attempt to pursue the logic of this argument is Phil Cohen's study of working-class subcultures which is grounded geographically in a specific locale (P. Cohen 1972). In this seminal paper Cohen argues that the traditional form of white working-class culture in East London was forged within the social context of the extended family and within the ecological setting of the working class

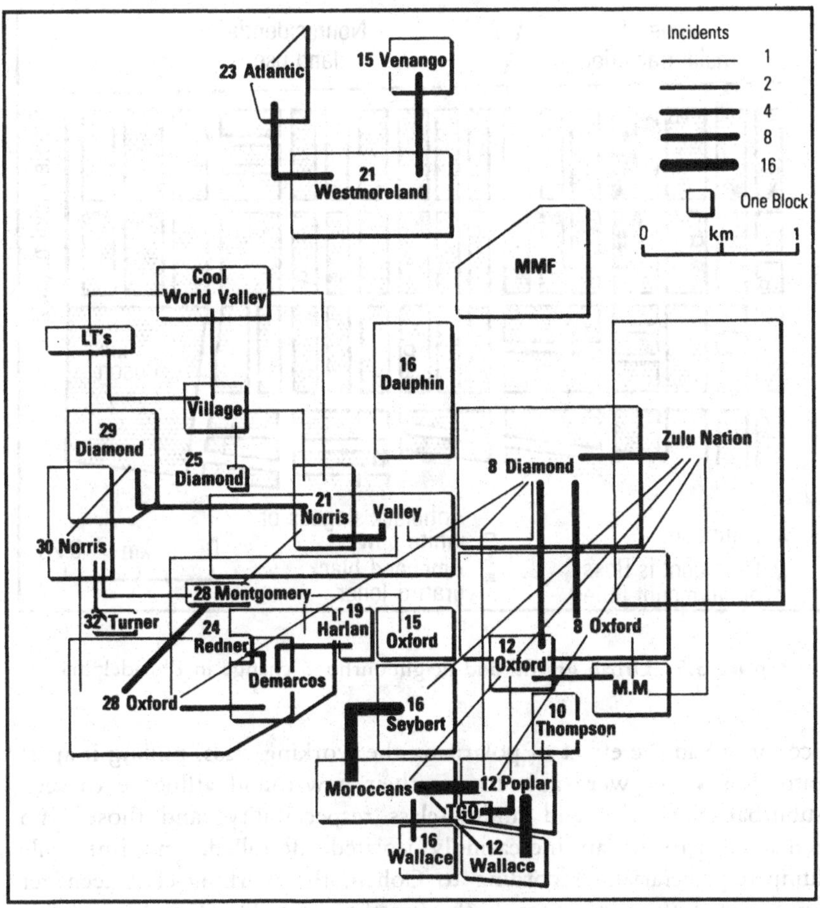

Figure 3.4 Gang violence in Philadelphia, 1966–70

neighbourhood. The local economy was once far more diversified than it is today, although people have always tended to live and work locally. Postwar redevelopment had catastrophic effects on the traditional pattern of working-class life, leading to the breakup of neighbourhoods through wholesale depopulation, the fragmentation of extended families, the influx of immigrant labour, and the decimation of local employment opportunities. The replacement of terraced houses by high-density, high-rise council estates destroyed the function of the street, the local pub, the corner-shop, and the informal articulation of communal space. The labour force became increasingly polarized into a privileged sector associated with the new technology and an unskilled sector of labour-intensive, low-paid, dead-end jobs.

Cohen argues that these structural changes in the local urban

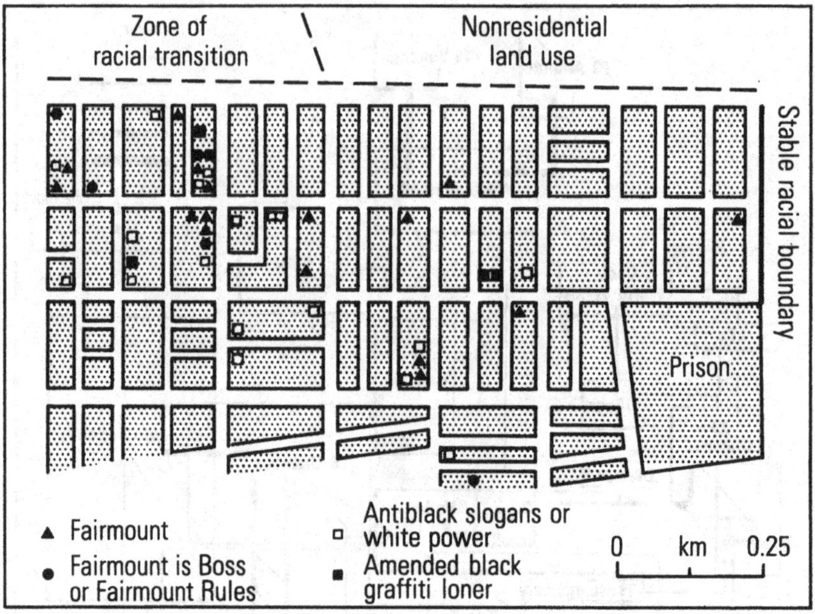

Figure 3.5 Urban graffiti and neighbourhood change in Philadelphia

economy had the effect of polarizing the working class, pulling it apart into those who were able to use their new-found affluence to seek suburban life-styles and middle-class respectability, and those who remained part of an increasingly isolated, deskilled, and immobile lumpen proletariat. According to Cohen, the working-class teenager experienced these changes both directly, in political and economic terms, and indirectly, in cultural terms. The material and ideological dimensions of social change intersected to produce certain determinate local effects. The youth subcultures that developed at this time can therefore be interpreted as having effectively resolved the contradictions of structural change in a symbolic or 'imaginary' way:

> Mods, Parkers, Skinheads, Crombies are a succession of sub-cultures which all correspond to the same parent culture and which attempt to work out through a series of transformations, the basic problematic or contradiction which is inserted in the sub-culture by the parent culture. So you can distinguish three levels in the analysis of sub-cultures: one is the historical . . . which isolates the specific problematic of a particular class fraction . . . secondly . . . the sub-systems . . . and the actual transformations they undergo from one sub-cultural moment to another . . . thirdly . . . the way

> the sub-culture is actually lived out by those who are [its] bearers and supporters (P. Cohen 1972; quoted in Clarke *et al.* 1976, p.33).

Cohen's argument is represented diagrammatically in Figure 3.6. It serves as an exemplary exercise in tracing the specific ways in which general shifts in class relations (Cohen's 'determinate conditions') feed through into particular subcultural responses at the local neighbourhood level. Despite the inherently geographical flavour of this work its implications have not yet been seriously considered by social and cultural geographers.

Cheerleaders and ombudsmen?

This chapter has considered the ideological nature of culture as derived from Gramsci's reworking of the concept of hegemony. Cultural practices have ideological effects to the extent that they contribute to the domination of one social group by another through the selective concealment of interests. This process follows predictable lines, given a knowledge of the structure of inequalities that characterize any particular society. This chapter has concentrated on the nature of counter-hegemonic resistance through an analysis of (mainly working-class) subcultures. But it should be recognized that dominant cultures also demand further research. Using Parkin's theory of social closure, an argument has been advanced to explain why resistance is often restricted to the symbolic level. This chapter has also considered the nature of subcultural style and its attempted resolution of structural contradictions in 'imaginary' or symbolic terms. Finally, the chapter has indicated the potential significance of geographical research on the territorial dimension of local ideological struggles.

One final question that needs to be considered here concerns the place of social scientists in the relations of dominance and subordination that they seek to analyze. While geographers have been slow to abandon their self-appointed role as the 'translators' of working-class and ethnic subcultures, they have scarcely begun to consider the ethical problems that this kind of research involves. In studying subordinate cultures, for example, cultural geographers run the permanent risk of becoming uncritical 'cheerleaders' of subcultural resistance (Bourne & Sivanandan 1980). As academics, social scientists have all too often assumed the guise of visitors to the human zoo, going slumming for a while in order to report back to their peers and paymasters about conditions in some exotic corner of the 'real world'. In relinquishing the apparently impartial role of 'ombudsman', the alternative danger is of becoming an uncritical cheerleader, romanticizing subcultural forms however earnest

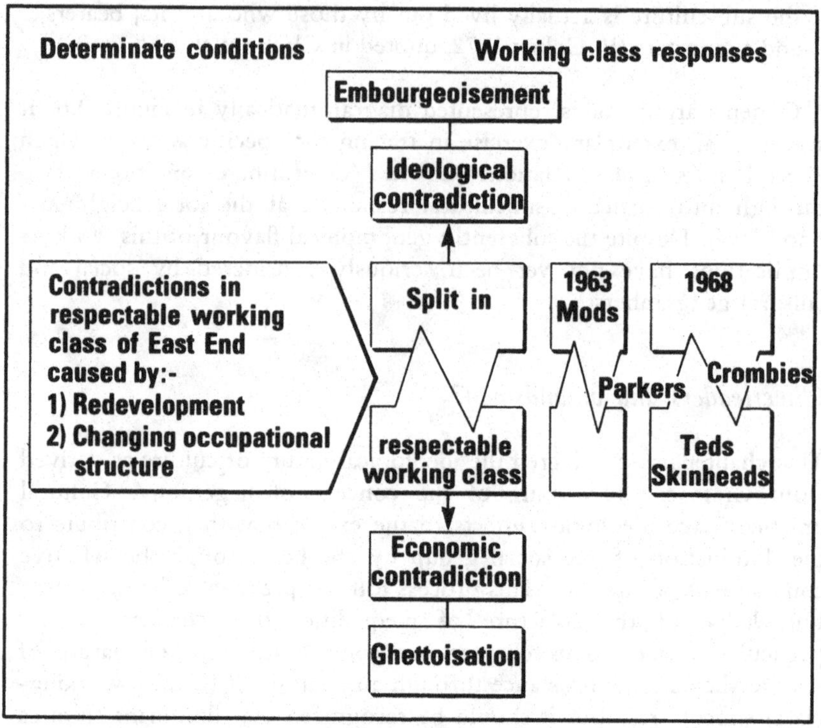

Figure 3.6 Structural conditions and working-class subcultures in East London

the effort to understand them. These twin dangers are best understood as a direct reflection of the structured inequality between academics and the communities they seek to study. The structural distance between the two groups creates the kind of vacuum in which ideology flourishes. Research strategies such as participant observation provide only an imperfect means of bridging this gap and resolving the difficulties that surround it.

The dangers of championing working-class 'resistance' and celebrating the authenticity of 'working-class culture' should not be underestimated. Middle-class academics frequently find the blatant racism and sexism of working-class life reprehensible, failing to recognize the extent to which the same issues pervade their own lives and the institutions in which they are employed. A partial answer to this particular dilemma can be found in the useful distinction between 'situated' and 'generalized' knowledge that allows people to hold stereotyped and derogatory opinions about *groups* of people while respecting and befriending *individuals* who are members of that group. But the general problem is not so easily resolved. The question of cultural relativism arises no less in the apparently 'familiar' setting of the contemporary city than in the apparently more 'exotic' context of traditional anthropology. At least some awareness of these debates, derived from anthropology (Barnes 1979; Cassell 1980), can only be of benefit to a revitalized cultural geography.

Notes

1 Antonio Gramsci (1891-1937) spent the last ten years of his life in prison under Mussolini. His fragmentary writings were eventually published as a six volume edition entitled the *Quaderni del carcere* (1948-51). English translations are avilable as *Selections from the prison notebooks* (1971) and *Letters from prison* (1973). A collection of Gramsci's cultural writings has also been published (Gramsci 1985).
2 The history of 'counter-hegemonic' struggles and their political significance has been traced in E.P. Thompson's classic study of *The making of the English working class* (1963).
3 Cynthia Cockburn's study of union organization and technological change in the printing industry is a brilliant analysis of the exclusionary practices employed by a group of working-class men to serve their own patriarchal interests (Cockburn 1983).
4 Stanley Cohen has objected to calling purely unconscious behaviour 'symbolic' or 'ritualized', arguing that symbolic behaviour requires a knowing subject, pursuing intentional action, or 'at least dimly aware of what the symbols are supposed to mean' (Cohen 1987, p.xiv).
5 Peter Bailey launches a similar attack on the sociology of leisure in his introduction to the paperback edition of *Leisure and class in Victorian England* where he warns of the dangers of over-politicizing leisure as an arena of struggle (Bailey 1987, p.9). Elsewhere, he has criticized cultural studies for its 'full frontal theoreticism, its often laboured conflation of the abstruse and the banal, and its tendency to overcomplicate' (Bailey 1986, p.xix).
6 An exception to the general dearth of ethnographic work is provided by Gary Armstrong who is completing a PhD in Anthropology at UCL, based on three years' participant observation with a group of Sheffield United supporters. Robins & Cohen (1978) also provide interesting ethnographic evidence from their work in London, showing, in the case of Arsenal fans, how different groups have quite different reasons for supporting their team.

Chapter four
Popular culture and the politics of class

> As an area of serious historical work, the study of popular culture is like the study of labour history and its institutions. To declare an interest in it is to correct a major imbalance, to mark a significant oversight. But, in the end, it yields most when it is seen in relation to a more general, a wider social history.
>
> (Stuart Hall 1981)

The meaning of 'popular culture'

If 'culture' is one of the most complicated words in the English language (Williams 1976), its qualification by the adjective 'popular' seems only to increase its ambiguity. Etymologically, the sense of the term 'popular' evolved from its literal meaning, 'belonging to the people', to its current implication of 'widely favoured' or 'well-liked' (as in 'popular music'). The association of popular taste with vulgarity, triviality, and baseness is now widespread, as indicated by the common distinction between élite and popular culture (Burgess & Gold 1985). Before going any further, therefore, it is worth tracing the variable meanings of this highly contentious term.

'Popular culture' has a very different political connotation from 'mass culture', the term preferred by the critical theorists of the 'Frankfurt School'.[1] For Theodor Adorno and his circle, the 'masses' were in danger of being debased by the endless diet of 'mass culture' to which they were subjected, rendering them easy prey to manipulation by an authoritarian state. 'Popular music', for example, was contrasted with 'serious music', the former being standardized, mechanized, and having a soporific effect on the social consciousness; the latter requiring effort, concentration, and high technical competence, disrupting the continuum of everyday life and encouraging recollection. In contrast to this pejorative view of the 'masses' and their debased tastes, 'popular culture' implies a much more positive evaluation of the people and their creative potential.[2]

In contemporary usage, popular culture is almost invariably set against élite culture ('the best that has been thought or known', in Matthew Arnold's celebrated phrase). Popular culture therefore has a definite political edge. Arnold's definition of (élite) culture was articulated in the context of his fears of imminent class conflict. Britain, he thought, faced a simple choice between culture and anarchy (Arnold 1869). As later critics like T.S. Eliot and F.R. Leavis reaffirmed, 'culture' (in Arnold's sense) is inherently undemocratic. Eliot went on to trace a moral geography around his élitist conception of culture (Cresswell 1988), arguing for a tight connection between people and place:

> Certainly an individual may develop the warmest devotion to a place in which he (*sic*) was not born, and to a community with which he has no ancestral ties. But I think we should agree that there would be something artificial, something a little too conscious, about a community of people with strong local feeling, all of whom had come from somewhere else. I think we should say that we must wait for a generation or two for a loyalty which the inhabitants had inherited (Eliot 1948, p.52).

Indeed, he continued, 'the majority of human beings should go on living in the place they were born', except presumably for the educated élite who, like Eliot himself, would continue to enjoy an exceptional degree of personal mobility.

Popular culture has therefore been championed mainly by those on the Left who, like E.P. Thompson, draw a distinction between 'plebeian culture' and 'patrician society' (Thompson 1974). Iain Chambers, for example, contrasts 'official culture' with 'popular culture', the one preserved in art galleries, museums, and university courses, demanding cultivated tastes and formally imparted knowledge; the other, more incidental, transitory, and expendable, not separated from daily life (Chambers 1986). Others define the tension in terms of sacred and profane, implying not just the opposition of popular culture to the 'sacred' culture of the élite, but also emphasizing its positive potential, celebrating 'the essential, rare, irreverent, gift of profanity: creativity' (Willis, 1978, p.170).

The meaning of 'popular culture' is further complicated by its historical associations with folklore and rural tradition. In the United States, this tradition is upheld by the Society for the North American

Cultural Survey, whose atlases chart the distribution of anything from folk music (fiddling styles) to urban deprivation (rat bites), though their preference is clearly for the vernacular and folksy (Rooney *et al.* 1982). Unlike much cultural geography, however, contemporary popular culture is predominantly urban in character, a reflection of the metropolitan experience (Chambers 1986). Moreover, 'popular' conveys an implication of *resistance* to conventional authority. But the popular can also be 'traditional', a tension that has made some writers dubious about using the concept at all (Hall 1981). For some writers, 'popular culture' is too inclusive: if it refers to 'the people' in general then it is hard to see what, if anything, can be excluded. For others, it is too exclusive, implying that there is little or no interaction between the cultures of the élite and the people.

Is 'popular culture', then, to be equated with the culture of the working class and, if so, how is class to be defined? It is all too easy to assume a homogeneity within the working class and to make a simple equation between social classes and cultural forms that cannot be sustained in practice (Clarke *et al.* 1979). Indeed, the literature on working-class culture often seems to assume that 'the people' are all young, male, and working class (cf. McRobbie & Garber 1975). Without labouring the point, it should be clear that the term cannot be used as if its meaning were self-evident, unambiguous, or uncontested. Yet it serves as a convenient label and a focus on some central issues of current concern in cultural studies about the way dominant meanings are contested by subordinate groups – a process, it will become clear, that is often inherently geographical.

This chapter focuses on the cultural patterns of the working classes in 19th-century British cities, with some cross-references to the United States. This period is of interest because it was one in which dominant meanings were particularly fiercely contested; a time of hegemonic crisis in the sense described in Chapter 3. Political upheaval and social turbulence were felt most strongly in the cities, themselves a reflection of the social geography of rapid industrialization (Dennis 1984). Under these conditions, 'social control' was by no means guaranteed. Indeed, some historians have suggested that the term 'social control' should be restricted to circumstances such as these (Thompson 1981, Donajgrodski 1977), while others argue that the term should not be employed at all.[3] Despite these difficulties, 'social control' has the virtue of directing attention towards the *relations between classes* rather than assuming that unitary, homogeneous 'class cultures' can be unproblematically identified. Before proceeding to an analysis of 19th-century popular culture, it is useful to establish what preceded it in the pre-industrial period, to outline what Laslett (1965) calls 'the world we have lost'.

Popular culture in pre-industrial Britain

It was during the period of transformation to an industrial society that many aspects of contemporary popular culture took on their current form. Before then, 'leisure' did not exist as a separate domain, readily distinguishable from the world of work. Other features of popular culture in urban industrial society appeared for the first time during the 19th century: annual holidays, for example, only became available to large sections of the working population after the Bank Holidays Act of 1871. Around this time, too, leisure became strictly segregated by social class: to take a holiday in Bournemouth was a decidedly different experience from the commoner pleasures of Blackpool. Important gender distinctions can also be traced to this period, although the concept of leisure has never been easy to apply in the case of women, for whom the category of 'work' has never been clear-cut (cf. Pahl 1988). Class and gender segregation were, however, a noticeable feature of British pub life where social distinctions between the lounge and the public bar were clearly demarcated (Mass Observation 1986). The case should not be over-stated, however, and a brief review of popular recreations in pre-industrial Britain shows that issues of 'social control' were not entirely a 19th-century innovation.

Peter Burke's work on the popular culture of early modern Europe draws on folklore, literary criticism, and social anthropology to uncover the myths, images, and rituals of the period 1500-1800. He argues that traditional forms of popular culture were threatened, even before the Industrial Revolution, by the growth of towns, the improvement of roads, and the spread of literacy (Burke 1978, p.6). Many forms of popular entertainment, such as Carnival, were participatory events in which all sections of society took part:

> Carnival . . . was for everyone. In Ferrara in the late fifteenth century, the Duke joined in the fun, going masked in the streets and entering private houses to dance with the ladies. In Florence Lorenzo de' Medici and Niccolo Machiavelli took part in carnival (ibid. p.25).

Although this example is useful in questioning the existence of a separate domain of 'popular culture', restricted to the masses and with no involvement from the élite, it fails to draw sufficient attention to the conflict that was latent in pre-industrial popular culture (Yeo & Yeo 1981). This deficiency is redressed elsewhere in Burke's analysis of the period where he draws on Max Gluckman's anthropological studies of the 'licence in ritual' (Gluckman 1956) to describe the sense of opposition implicit in the traditional symbolic reversals of Carnival and related festivals. A classic example is Pieter Brueghel's representation of *The Battle of Carnival and Lent* (1559), where Carnival is depicted as a jolly fat man, seated on a barrel, jousting with an emaciated woman who represents the material privations of Lent (Fig. 4.1). The left half of the painting is filled with the symbols of Carnival: food, drink, sex, and violence. The right half represents the religious and social restraint appropriate to Lent.

Throughout southern Europe too, Carnival was the great popular festival of the year, 'when what oft was thought could for once be expressed with relative impunity' (Burke 1978, p.182). As with Bakhtin's analysis of medieval Carnival (see Ch. 7), Burke stresses its spatial and temporal structure. Ritualized inversions of the social order were tolerated, even encouraged, because they were acknowledged by everyone to be a temporary respite from the conventional social order to which everything would return in due course. Carnival took place, literally, in a world apart, in the city centre and in the open air:

> Carnival may be seen as a huge play in which the main streets and squares become stages, the city became a theatre without walls and the inhabitants, the actors and spectators, observing the scene from their balconies. In fact, there was no sharp distinction between actors and spectators, since the ladies on their balconies might throw eggs at the crowd below, and the maskers were often licensed to burst into private homes (Burke 1978, p.182).

The tension in Carnival, then as now, is between 'social control' and social protest. Some have argued that festivals like Carnival serve as a safety valve, providing a relatively harmless and ritualized way for subordinate groups to express their sense of injustice, with order maintained by a diet of 'bread and circuses'. Keith Thomas comes close to adopting this position when he suggests that Saturnalia and similar festivals in pre-industrial Europe were an occasion for the kind of 'periodic release necessary in a rigidly hierarchical society' (Thomas 1964, p.53). The ritual trial and execution of King Carnival on Shrove Tuesday (Mardi Gras) can also be taken as a symbol of the resumption of normal social relations, marking the end of Carnival and the beginning of Lent.

Other historians maintain that the symbolic reversals of day and night, male and female, rich and poor, do more than simply reaffirm status differences. In Burke's own discussion of Carnival, both possibilities are admitted as people switch codes 'from the language of ritual to the language of rebellion' (Burke 1978, p.203). Riots frequently broke out on the occasion of major street festivals, for example. But it is not simply the case that the cultural occasionally flips over into the

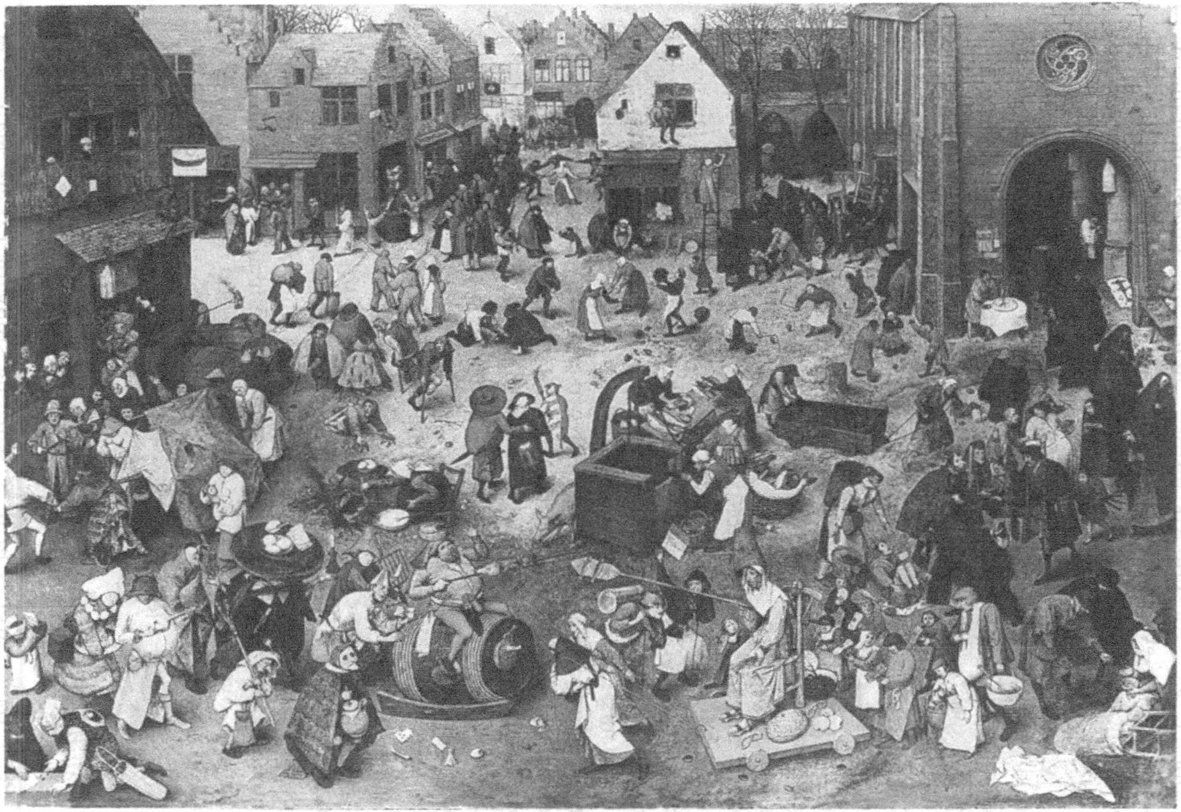

Figure 4.1 Breughel's *The Battle of Carnival and Lent* (1559)

political. This implies an artificial distinction between culture and politics, coupled with that E.P. Thompson (1971, p.76) calls 'a spasmodic view of popular history'. With the advent of industrial society, similar events presented even more of a threat to the established social order, not least because of the increased density and heterogeneity of urban life.

The 19th century saw major changes in popular recreation as the world of leisure became increasingly divorced from the world of work and as an industrial working class began to experience leisure as a separate sphere of life. The temporal aspects of this transformation have been well documented, in E.P. Thompson's brilliant essay on time and work-discipline under industrial capitalism (Thompson 1967) among other sources. While the celebration of 'Saint Monday' remained a popular means of resistance to the increasingly synchronized and regulated demands of manufacturing, the spatial constitution of popular resistance has been much less well observed. There was, however, a distinct social geography to 19th-century popular culture. It was not simply that work and leisure became separate temporal domains, but that leisure time in the industrial city came to be spent outside the workplace, beyond the scrutiny of factory owners and managers. Early experiments in town planning, such as Saltaire (1853), Bournville (1893), and Port Sunlight (1888), represent clear attempts to translate the paternalistic impulse of an earlier generation into a form that was more suitable to the Victorian city (see Fig. 4.2). Philanthropic housing trusts also tried to impose a similar 'moral order', providing subsidized housing for the 'deserving poor'.

Rather than proliferate these examples, two particular cases have been selected to demonstrate how the control of space was as important as the control of time in the imposition of work-discipline and the establishment of a bourgeois hegemony: the battle of the music halls and the battle of the streets. Both cases support David Harvey's contention that 'command over money, command over space, and command over time form independent but interlocking sources of social power' (1985a, p.1).

Fun without vulgarity: the battle of the music halls

As Peter Bailey argues in his fascinating study of *Leisure and class in Victorian England* (1978), the paternalistic tolerance of the upper classes in pre-industrial England gave way to a sour impatience towards plebeian culture with the onset of industrialism. To the Victorian bourgeoisie, popular culture often appeared to be morally offensive,

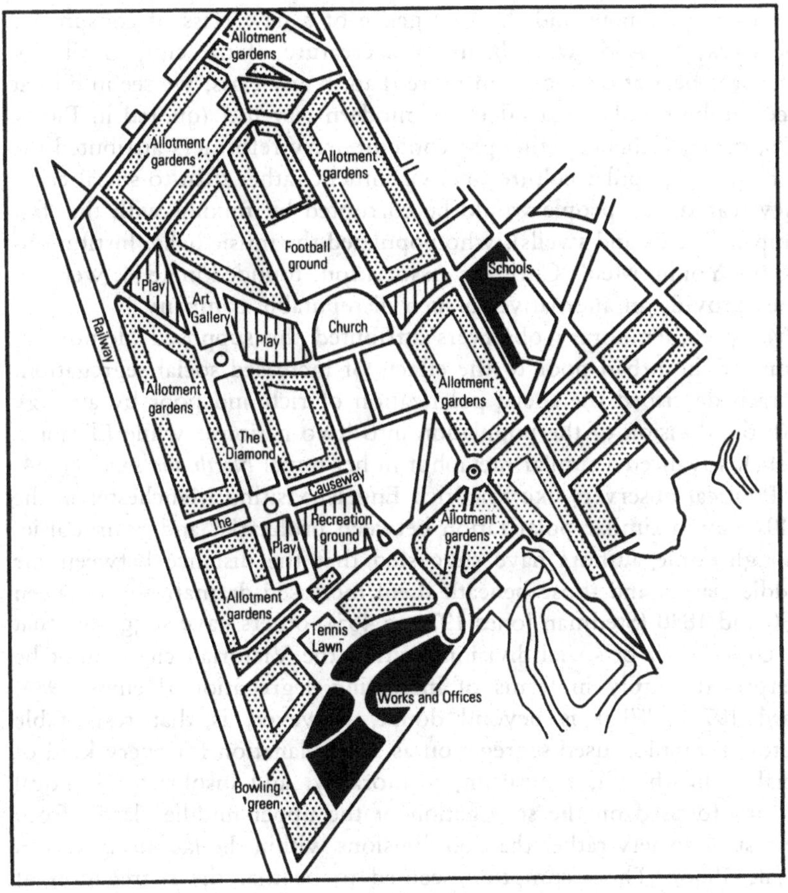

Figure 4.2 Plan for Port Sunlight, 1917

socially subversive, and a general impediment to progress. A social reform movement, calling for the development of more 'rational recreation', was launched, addressing itself to the regulation of working-men's clubs, to the encouragement of athleticism, and, above all, to curbing the perceived excesses of the music halls.

Average wages rose steeply in the Victorian city during the latter half of the 19th century and especially during the 1870s. Much of this money was spent on leisure and entrepreneurs were quick to exploit the expanding urban market for popular recreation, cashing in on the 'business of pleasure' (Bailey 1986). Charles Kingsley, author of *The Water Babies* (1863) and an active proponent of 'rational recreation' and 'muscular Christianity', saw the link between the growing demand for urban entertainment and the emergence of a new class of consumers. The clerk, he said, was 'distinctly a creature of the city; as all city influences bear at once on him more than in any class, we see in him at once the best and worst effects of modern city life' (quoted in Bailey 1978, p.93). Generally, though, contemporary reformers attributed the problems of popular culture to environment rather than to social class. They feared that people would be corrupted by mixing with the 'fast company' (cads and swells), who populated the music halls. Institutions like the Young Men's Christian Association, founded in 1844, were set up to provide an alternative to such disreputable company.[4]

Many contemporary observers attributed the supposed 'demoralization' of the urban poor to the effects of increased social segregation. Disraeli described the social polarization of rich and poor by analogy with the division of the population into 'two nations', while Elizabeth Gaskell advanced a similar metaphor in her novel *North and south* (1854-5). Political observers like Friedrich Engels, visiting Manchester in the 1840s, saw a similar polarization between proletariat and bourgeoisie. Though some authors have suggested that the distance between the middle classes and those beneath them increased dramatically between 1790 and 1840 (Stedman Jones 1974, p.465), others have suggested that the complexity of social divisions within the Victorian city cannot be interpreted simply in terms of increasing segregation (Dennis 1984, Ward 1976). What is beyond doubt, however, is that respectable Victorian opinion used segregation as an explanation for every kind of social evil: atheism, radicalism, immorality, and insobriety. Political concern focused on the segregation of the upper-middle classes from the rest of society rather than on divisions within the labouring classes (Neale 1968). The assumption seemed to be that the segregation of upper and lower classes would result in the corruption of good 'yeoman' English stock by the contagious example of the Irish and other 'undesirables' who were then crowding into the towns. They feared the spread of disease from the insanitary conditions that prevailed in the cities and looked back nostalgically to the idealized 'community' of the country mill-town, which provided the model for the utopian visions of industrialists like Titus Salt and George Cadbury (Fig. 4.3).

In a situation of increased social segregation, popular recreation provided one of the few arenas in which different social classes occasionally came into contact. The resulting tensions were entirely predictable. Public excursions to the countryside, for example, which grew in popularity during the 19th century, were greeted with abhorrence by outraged middle-class observers. In the following childhood reminiscence, the novelist, Ouida, expresses her repugnance at the descent of the 'townie' on rural Derbyshire:

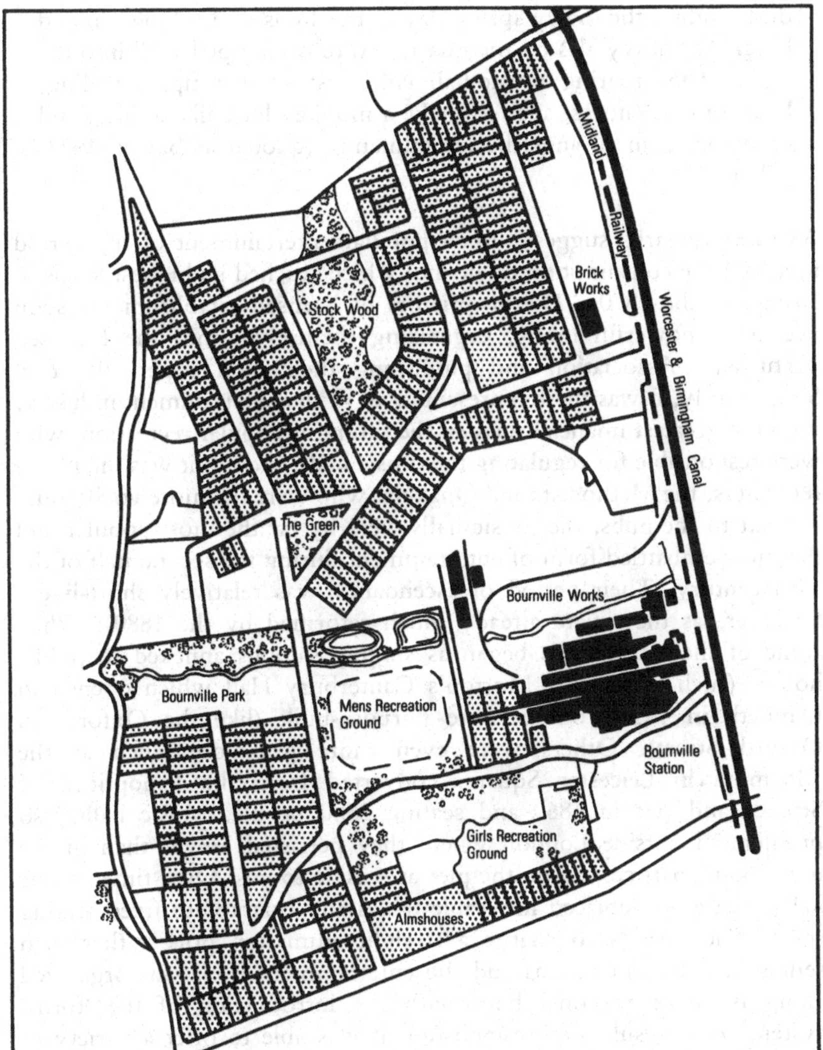

Figure 4.3 Plan for Bournville, 1898

The excursions trains used to vomit forth, at Easter and Whitsun week, throngs of millhands of the period, cads and their flames, tawdry, blowzy, noisy, drunken; the women with dress that aped 'the fashion', and pyramids of artificial flowers on their heads; the men as grotesque and hideous in their own way; tearing through the woods and fields like swarms of devastating locusts, and dragging the fern and hawthorn boughs they had torn down in the

> dust, ending the lovely spring day in pot-houses, drinking gin and bitters, or heavy ales by the quart, and tumbling pell-mell into the night train, roaring music-hall choruses; sodden tipsy, yelling, loathsome creatures, such as make a monkey look like a king, and the newt seem an angel beside humanity (quoted in Bailey, 1978, p. 104).

As these remarks suggest, much popular entertainment of the period involved the consumption of alcohol which reached its highest levels in Britain in the 1870s. The proprietors of licensed establishments soon became a powerful lobby, organizing themselves into the Licensed Victuallers' Association and publishing their own paper, the *Era*. Significantly, it was these representatives of the entertainment industry, with the greatest financial stake in the future of popular recreation, who were responsible for regulating the music halls, though it was the moral reformers, the Methodists and Quakers, who were the more vociferous.

Next to the pubs, the music halls were by far the most popular and the most embattled form of entertainment during the second half of the 19th century. Their period of ascendancy was relatively short-lived, however, as they were already much reformed by the 1880s. While some of the music halls began as singing saloons annexed to public houses (such as Charles Morton's Canterbury Hall which opened in Lambeth in 1851), others were purpose-built (like the Oxford, in Oxford Street). Others were even more grandiose, such as the Alhambra in Leicester Square, converted from the Panopticon of Science and Art in 1860 and seating 3500 people. Music halls also proliferated outside London where they persisted longer than in the metropolis, particularly on the pier at seaside towns. Over time, music hall proprietors replaced moveable tables and chairs with fixed seating stalls. They began to charge a straight admission price rather than selling refreshment tokens and the entertainment itself was organized along more professional lines with the introduction of the 'turns' system. As a result, each establishment was able to offer a variety of acts, with individual performers travelling across the city to do their 'turn' in a number of different venues in the course of a single evening.

With two or three hundred halls in London alone during the 1850s, competition was fierce, innovation intense, and the pace of change extremely rapid. Relatively little information exists about the content and style of particular performances[5], but the atmosphere of the halls seems to have been one of gregarious congeniality. The audience was free to smoke and drink, eat and talk, even at the height of the performance. Unlike so-called legitimate theatre, where there was a clear distinction between audience and actors, music hall audiences engaged in an active dialogue with the players. They expressed their approval or disapproval with gusto, joining in the choruses of songs or pelting the performers with whatever they had to hand according to their mood. Despite a sprinkling of nobility and upper-class patrons, including the Prince of Wales, music hall audiences were predominantly composed of artisans, tradespeople, and clerks. To respectable members of the bourgeoisie, the music halls personified the debased taste of the masses, their characteristic drunkenness and debauchery, and the potentially riotous nature of unregulated working-class recreations.

These social tensions had a distinctive geography which expressed itself at various levels. Much of the offensiveness of the music halls to respectable values arose from their public prominence within the city:

> Built on the main thoroughfares and emblazoned with posters, they ranked second only to the new town halls in size and capacity as places of indoor assembly. Music hall advertising was ubiquitous (Bailey 1978, p.8).

But there was also a distinctive social geography inside the halls reflected in the price and arrangement of the seating. The following account describes the arrangement of a typical 19th-century music hall audience in the United States, where 'each section of the theater was a society of its own':

> . . . various sections of the theater attracted different social and economic groups. The expensive box seats offered privacy, prestige, the greatest measure of decorum, and a place for proper ladies to sit . . . Below the boxes, in front of the stage, was the pit, where the 'middling' classes sat. Depending on the circumstances and the disposition of the commentator, the people in the pit were both attacked as crude ruffians and praised as good people interested in the productions. But there was no question about the nature of the gallery, which was located in the upper reaches of the theater and occupied by what were thought to be the lowest reaches of society. If Negroes were allowed ino the theater at all, the management confined them to part of the gallery . . . It was from the gallery that missiles of all sorts rained down on unpopular performances or in response to unpopular material. Social whispering might sweep the boxes, but loud bellows rang from the gallery patrons who actively and vociferously participated in the performances (Toll 1974, p.10).

Performances were also commonly divided into two parts: the first half devoted to more refined material; the second including more profanity and 'low' humour. Temporal and social segregation was reinforced by segregation according to seat price: the Canterbury, for example, charged sixpence for admission to the pit and ninepence for the gallery.

The halls were accused of every sort of impropriety. They were held responsible for encouraging alcoholism, poverty, class conflict, and moral degeneracy ('ante-rooms to the brothels'). Reform was effected by regulating the halls through the annual renewal of music and drinking licences, and by regulating the content of the material that was performed. Many of the songs involved thinly disguised sexual innuendo (such as the 'swell') or political satire. Neither was considered appropriate material for 'rational recreation' and both became the target of censorship. Rather than risk losing their licences, the halls began to impose their own form of self-censorship by adopting a system of 'house rules'. The following example from 1883 is typical:

> Any artiste giving expression to any vulgarity, in words or actions, when on stage, will be subject to instant dismissal, and shall forfeit any salary that may be due for the current week (quoted in Bailey 1978, p. 165),

or again, from 1892:

> No offensive allusions to be made to any Member of the Royal Family; Members of Parliament, German Princes, police authorities, or any member thereof, the London County Council, or any member of that body; no allusion whatever to religion, or any religious sect; no allusion to the administration of the law of the country (ibid. p. 165).

Strictly enforced, these rules would have left scarcely any subject for humour and performers developed a number of strategies for evading the censors such as ad-libbing and suggestive intonation. The halls paid informants to catch those who over-stepped the mark, but 'house rules' seem mainly to have been for the protection of proprietors rather than for the control of performers. The very fact that managers felt obliged to post such notices is an indication of the strength of the reform movement and its translation into a policy of self-regulation. The same impulse lay behind Moore and Burgess' description of their shows as 'fun without vulgarity', a tag which neatly catches the way that music hall proprietors, if not their performers, internalized the Victorian ideal of respectability (Mair 1986). By the 1880s, the reform of the music halls had proceeded apace, extending to a ban on the sale of alcohol in the halls, leading one observer to quip: 'Abandon Hops All Ye Who Enter Here'.

The 'battle of the music halls' may have been a relatively easy victory for the forces of reform (in association with the entrepreneurs whose interest lay in keeping the halls open for business). But the threat they posed to bourgeois values was easily contained within the music halls themselves which provided the scope for effective management. The 'battle of the streets', whether for popular entertainment or more nefarious purposes, was a much more difficult and protracted affair. The lines between the various protagonists were much less clearly drawn and the control of public space was much more easily resisted than what went on indoors elsewhere.

Policing the public

In the second half of the 19th century, many bourgeois Victorians felt that 'civilization' was under threat from the 'urban masses'. The casual poor of 'outcast London' and other cities could not be ignored: their presence on the streets was a constant visual reminder of a social order on the point of crisis. In a classic study of relations between these classes in Victorian London, Gareth Stedman Jones draws attention to the geographical basis of the 'moral panic' that ensued, leading to a whole round of reform and regulation:

> In the course of the nineteenth century, the social distance between rich and poor expressed itself in an ever sharper geographical segregation of the city. Merchants and employers no longer lived above their place of work. The old methods of social control based on the model of the squire, the parson, face to face relations, deference, and paternalism, found less and less reflection in the urban reality. Vast tracts of working-class housing were left to themselves, virtually bereft of any contact with authority except in the form of the policeman or the bailiff (Stedman Jones 1971, pp.13-14).

Under these circumstances of growing social distance and increasing spatial segregation, the rich were free to entertain extravagant notions about the depravity of the poor whose humanity was quickly reduced to categories like the 'dangerous classes', the 'submerged tenth', and the 'great unwashed'.[6] Working-class districts were likened to immense *terrae incognitae* which were periodically visited by intrepid explorers and zealous missionaries from more prosperous parts of the city. The great social surveys of Henry Mayhew, Charles Booth, and Edwin Chadwick provided detailed social maps of this dangerous and hitherto uncharted territory. The metaphor is entirely appropriate: contemporary accounts of the Victorian city are full of the imagery of Empire. William Booth, founder of the Salvation Army and the first to refer to the 'submerged tenth', entitled his work *In darkest England and the way out* (1890), asking rhetorically:

> As there is a darkest Africa is there not also a darkest England? . . . ,May we not find a parallel at our own doors and discover within a stone's throw of our cathedrals and palaces similar horrors to those which Stanley has found existing in the great Equatorial forest? (quoted in Keating 1976, p.145).

The following year, the Reverend Osborne Jay chose a similar title: *Life in darkest London* (1891). Indeed, the impact of colonial affairs such as the Morant Bay rebellion in Jamaica in 1865 was debated with such intensity in England, not least because of the fears it reflected of a similar uprising at home.[7]

The metaphor of a 'dark continent' of poverty within the heart of British cities can, in fact, be readily extended to describe the policing of working-class neighbourhoods which was legitimized through popular notions of the 'natural' depravity of the working classes. Thus, in *Life and labour of the people* (1889), Charles Booth described the 'lowest class of occasional labourers, loafers, and semi-criminals' who were living 'the life of savages':

> They render no useful service, they create no wealth: more often they destroy it. They degrade whatever they touch, and as individuals are perhaps incapable of improvement; they may be to some extent a necessary evil in every large city, but their numbers will be affected by the economical condition of the classes above them, and the discretion of 'the charitable world'; their way of life by the pressure of police supervision (quoted in Keating 1976, p.114).

In the latter half of the 19th century, the police were commonly regarded as 'domestic missionaries' (Storch 1976), charged with the surveillance and control of the streets and other public places. In turn, the police were regarded as a pestilence by those they sought to control: 'a plague of blue locusts' (Storch 1975) resented particularly for the challenge they posed to the right of free public assembly. As the state attempted to extend its moral and political authority into areas of society that were previously regarded as of only marginal significance, the presence of the police was resolutely, sometimes violently, resisted.

With growing social tensions, particularly in urban areas, there was a great surge of interest among polite society in the condition of the urban poor. In the United States, books like Jacob Riis' *How the other half lives* (1890) attracted a massive audience who pored over the dramatically posed photographs of ragged street urchins and other scenes of urban poverty that illustrated his text (Fig. 4.4). The *New York Sun* published some of Riis' more picturesque photographs and *Scribner's* magazine carried a summary of the text. While Riis pioneered the art of photo-journalism, others like Octavia Hill indulged their taste for philanthropic 'good works', sanctimoniously averring their belief in the essential goodness of the people and responding avidly to the moral obligation they felt this placed on them. Their solutions combined pious motivation with intense practicality, as the following quotation from Octavia Hill, written in 1883, confirms:

> I always believe in people being improveable; they will not be improveable without a good deal of moral force, as well as improved dwellings; if you move the people, they carry the seeds of evil away with them (quoted in Stedman Jones 1971, p.193).

This combination of morality and practicality resulted in a variety of philanthropic endeavours in Britain and the United States, including the Settlement House Movement led by Jane Addams in Chicago and the Peabody Trust in London, set up by American banker, George Peabody, to provide subsidized housing for 'the deserving poor'.

More ambitious in its proposals for moral and social reform were the efforts of General William Booth and the Salvation Army to provide 'work for all' as a way out of the depravity and desperation of 'darkest England'. As the frontispiece to Booth's volume shows, Booth saw the problem of urban poverty, and hence its solution, in fundamentally geographical terms (Fig. 4.5). The Salvation Army would pull people out of the 'sea of starvation' and into the 'City Colony'. Then, for those who proved themselves worthy of further assistance, a 'Farm Colony' would be established at some distance from the town. Finally, emigration to the British Colonies, Foreign Lands and the (yet to be established) Colony Across the Sea was also contemplated. There is a wealth of meaning in Booth's social map, not least in terms of the way that social polarities (between crime, drink, and shame on the left hand column and destitution, despair, and death, on the right) are mirrored geographically in the tensions between town and country, home and abroad. The City Colony contains an idealized world of suburban villages twelve miles from town, salvation factories, food depots, and the prospect of permanent work in the provinces. Whitechapel, in East London, is magically relocated 'by-the-sea', as co-operative farms and smaller allotments (three acres and a cow) are established in the Farm

Figure 4.4 'Street Arabs' in Mulberry Street, New York

Colony. This comprehensive moral vision includes a fascinating tension between the almost obsessive interest in morbid detail (2297 suicides and 2157 found dead last year) and the much more abstract, qualitative concern for concepts like uncleanness, covetousness, and lack of righteousness.

In America, even more extreme measures were considered than Booth's emigration proposals. Reformers blithely recommended the enforced separation of children from their parents if the parents' circumstances were judged to be sufficiently corrupting.[8] Likewise, Booth's implication that 'rewards' had to be earned was by no means unique. Even the most nobly motivated philanthropic gesture had its price. Every gift involved an obligation: in order to receive one had to behave in an acceptable manner, returning gratitude and humility for the charity received.

As in contemporary Britain, the distinction between popular culture and public disorder in Victorian cities was not always very clear. The rash of public disorders that broke out in the latter half of the 19th century were interpreted as evidence of the demoralization of the poor rather than as an indication of the desperation of their circumstances. A programme of slum clearance was enacted to remove the so-called 'rookeries' of central London, condemned as the breeding ground of disease and political radicalism: the foci of 'cholera, crime and Chartism' (Stedman Jones 1971). There were riots in St. George's-in-the-East in London in 1859; 'bread riots' throughout the country in the winters of 1855, 1861, and 1866; and during February 1886 the Hyde Park riots shook the capital to its core. The windows of gentlemen's clubs were smashed along St. James's Street as groups of 'roughs' marched on Trafalgar Square and for a couple of hours the West End was in the hands of the mob. The poor seized the opportunity to settle accounts with the rich while the normal rules of social restraint were temporarily suspended. The seriousness with which the riots were taken can be seen from the immediacy of the reaction to them:

> The Mansion House Fund for the unemployed rocketed overnight. Fresh fuel was added to the flames by unemployed demonstrations in Birmingham, Norwich and other centres, and rioting in Leicester. The authorities in Glasgow found work for 895 unemployed in *one day* when the news of the Trafalgar Square riots came through. The middle and upper classes throughout the country reacted as if they had suddenly discovered a foreign army camping in their midst (Thompson 1955, p.407).

In March 1866, the socialist writer and artist William Morris felt he was witnessing the first skirmishes of the revolution. The force with which the subsequent riots were put down suggests that the authorities may have had a similar impression. In November 1887, during the disturbances of Bloody Sunday, the police made repeated baton charges on peaceful demonstrations by the Social-Democratic Federation (S-DF) and other Radical groups including the Irish National League. The demonstration was brutally suppressed. The Riot Act was read and a regiment of Guards with fixed bayonets was employed in an attempt to restore public order. While *The Times* leader writers condemned 'these howling roughs' and criticized their alleged 'love of disorder, hope of plunder, and the revolt of dull brutality against the rule of law' (14 November 1887), other people lamented the death of unarmed protestors at the hands of the police who had shown what E.P. Thompson called 'the true face of reaction'.

The public funeral of Alfred Linnell, a member of the S-DF who died in the riot, was attended by a large body of Radicals, Irish, and Socialists. The crowd sang a 'Death Song' written by William Morris, with a memorial design by Walter Crane showing a mounted policeman charging into a crowd of people carrying banners proclaiming justice and liberty (Fig. 4.6). The co-operation of Morris and Crane

IN DARKEST ENGLAND, AND THE WAY OUT.

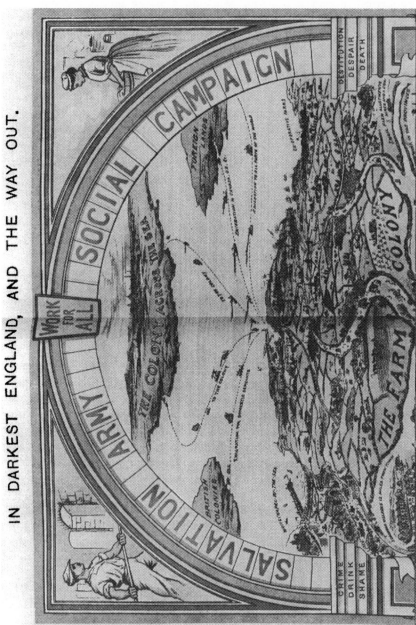

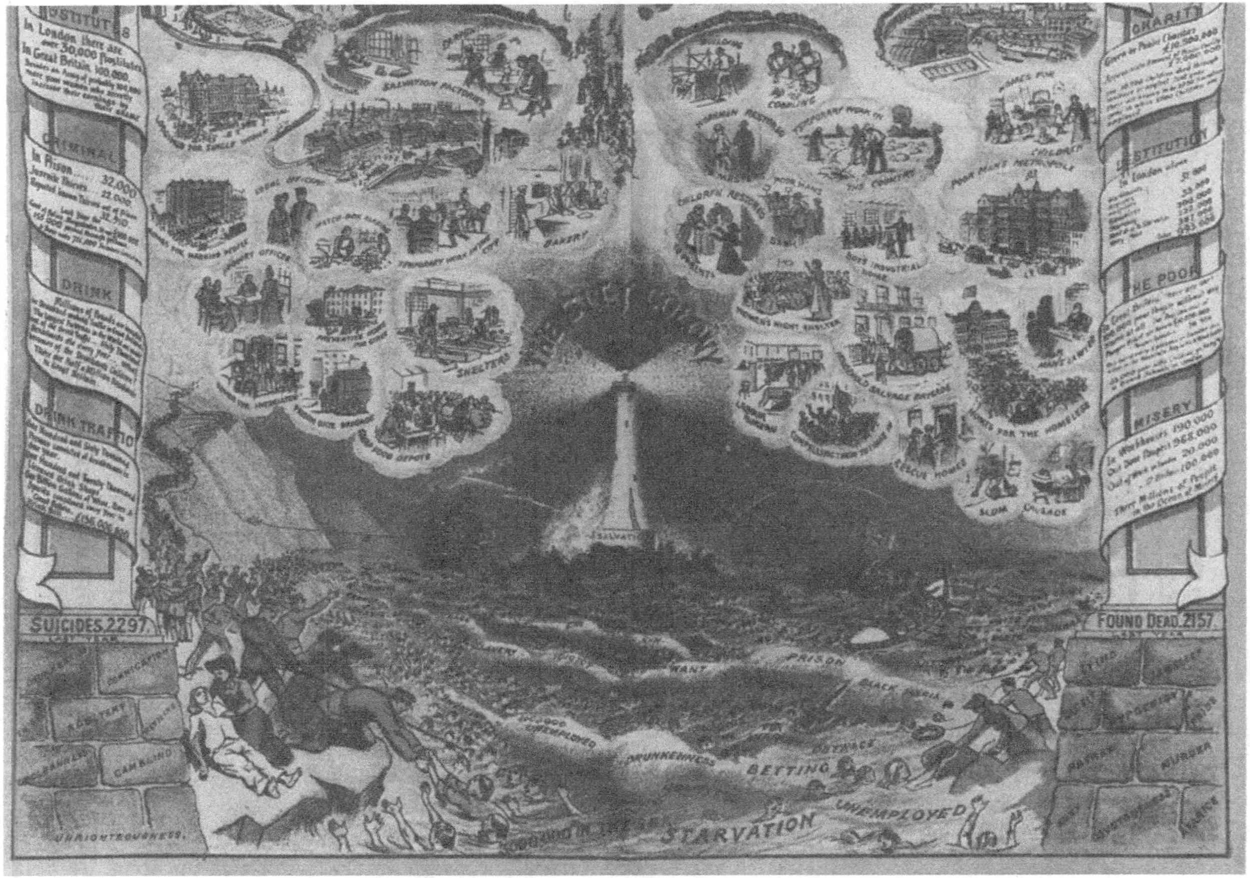

Figure 4.5 Frontispiece from Booth's *In Darkest England*. . . 1890

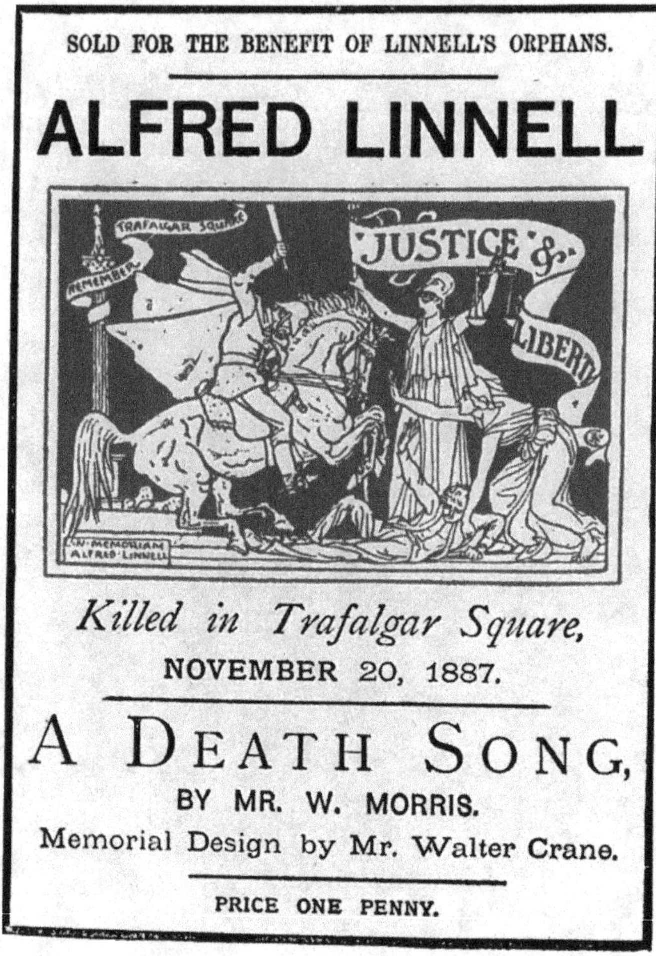

Figure 4.6 A death song for Alfred Linnell

in this venture demonstrates the depth of feeling that Linnell's death caused, producing an effective association between culture and politics. But why were these popular protests so forcefully put down? Why was the control of the streets such a central issue for the Victorian authorities?

For members of the élite, the 1880s were a period of intense social crisis, reflected in their hardened attitude towards the policing of the urban poor. Theories of 'demoralization' gave way to speculation about 'degeneration' as pauperism was replaced by chronic poverty. Paul Boyer describes a similar process in the United States, where the social reformers began to look to the environment rather than to individual morality as the cause of urban poverty (Boyer 1978). The British experience of a 'dark continent' of poverty at the urban core was also paralleled in the United States. As the Reverend Walter Rauschenbusch noted, not without irony:

> We have a new literature of exploration. Darkest Africa and the polar regions are becoming familiar; but we now have intrepid men and women who plunge for a time into the life of the lower classes and return to write books about this unknown race (quoted in Boyer 1978, p.127).

According to Boyer, the Haymarket riot of 1886 raised the 'triple menace' of class warfare, alien radicalism, and mass violence. The Pullman strike and the dispute at Pittsburgh Steel both ended in violent conflict, leading to similarly repressive measures as those employed in the Hyde Park riots in London. The social geographer, David Ward, has developed the comparison between British and American responses to urban slums at the turn of the century (Ward 1984). He notes the way that the reformers' imagery changed from 'natural' metaphors of abyss, flood, submergence, and decay to a medical and scientific vocabulary of pathology, plague, and epidemic, culminating in the ecological terminology of the Chicago School.[9] Moreover, in both Britain and America the issue of 'social control' was concentrated on specific spatial domains: the brothel, the saloon and, especially, the streets.

The Victorians' preoccupation with prostitution was one area where questions of private morality and public behaviour came into particularly sharp focus. Resolution of these questions often involved an interplay between 'high' and 'popular' culture, a striking example being William Holman Hunt's *The Awakening Conscience* (1853), in which a 'fallen woman' remorsefully recalls her childhood home in the course of singing a popular melody with her lover (Fig. 4.7). Holman Hunt painted the picture, which he considered the material counterpart of his most famous religious painting, *The Light of the World* (1853-6), after reading about the plight of Peggotty and Emily in *David Copperfield*. The painting is almost ludicrously iconographic: a cramped interior contrasted with the open air seen through an open window, reflected in a mirror; a cat devouring a bird beneath the table; the picture above the piano of the 'Woman taken in Adultery'. Even the wallpaper is symbolic: 'The corn and vine are left unguarded by the slumbering cupid watchers, and the fruit is left to be preyed on by thievish birds' (Holman Hunt, quoted in Hilton 1970, p.92).

Figure 4.7 William Holman Hunt's *The Awakening Conscience, 1853*

Prostitution was a key area in which the Victorian ideal of bourgeois respectability was defined and contested, demanding considerable ideological work to reconcile the sexual double-standard of middle-class men's routine recourse to (working-class) female prostitutes while at the same time upholding the sanctity of the family. Although there was a clear class basis to the sexual and financial transactions involved, the resolution of conflicting gender ideals took place largely in the cultural domain. Significantly, too, this contradiction and its resolution was expressed geographically.

As Christine Stansell shows in her superb study of New York, the Victorians' moral panic about prostitution reflected changes in the social and spatial relations of the trade, as well as the changing gender relations that were taking place in 19th-century cities:

> What disturbed observers was not just the number of women who bargained with men for sex, but the identity of those women. Moreoever, the entire context of the transaction was changing, as prostitution moved out of bawdy houses of the poor into cosmopolitan public spaces like Broadway . . . Prostitution was becoming urbane. The trade was quite public in the business district as well as in poor neighborhoods, a noticeable feature of the ordinary city landscape (Stansell 1986, p. 173).

Stansell argues that prostitution was not uniformly a tragic fate for every woman involved in the trade, nor was it invariably an act of defiance. Rather, for many women it was simply a way of getting by in an inherently unequal, patriarchal society. This attitude was itself an affront to Victorian respectability as it offered women a relatively independent income, decreasing their economic dependence on men (pimps only emerging in the early 20th century in New York). Prostitution was offensive above all, however, because it was so public. Women were regarded as the moral guardians of their children and were obliged to maintain strict standards of piety, decorum, and virtue in the home. The domestic sphere represented a social and spatial domain to which women were consigned by prevailing gender relations, but within which they could exercise a degree of control. The presence of women and children beyond the confines of the home, especially when they ventured into the public domain of the streets, represented an immediate affront to bourgeois morality. Prostitution was abhorrent not only in providing women with an (admittedly precarious) existence outside the domain of family and domesticity, but also because it achieved this by bringing the erotic into public space (ibid. 1986, p. 184).

Contemporary social reformers therefore tended to see the solution to the problem in spatial terms: 'urban social geography, not the landscape of the soul, engaged their ardor for exploration' (ibid. p.194). Tenement houses were regarded as the 'parent' of disorder and the 'nursery' of vice. The presence of children on the streets was taken as *prima facie* evidence of social pathology; not as a symptom but as a cause of poverty and corruption:

> The presence of children on the streets, besides being morally and epidemiologically dangerous, was proof of how tragically lacking the working poor were in this respect. From both standpoints, a particular geography of sociability – the engagement of the poor in street life rather than in the home – became itself evidence of a pervasive urban pathology (ibid. pp.202–3).

The street was symbolically opposed to the home: a profane versus a sacred world. The moral geography of family life was such that social reform took an inevitably spatial form: slum clearance, truancy and vagrancy laws, philanthropic housing, or the enforced removal of children from their corrupt parental home. A clear association was assumed between the private virtues of family life in the home and the public dangers of the streets. Moral order in the Victorian city was underpinned by its social geography (the segregation of classes and the separation of home and work, with its implicit gender division of labour); while 'social control' was maintained and contested through a series of distinctly spatial strategies.

Conclusion

To many middle-class Victorians, the city was a place of vice and immorality. As Britain became the first 'urban nation' (with more than 50 per cent of its population living in towns in 1851), new challenges were posed to the maintenance of public order. This chapter has examined how 'social control' was maintained and resisted in the Victorian city; how the politics of class were conducted through popular culture and the policing of the streets. Public and private domains interpenetrated in new and challenging ways, raising new tensions for the relations between social classes. Often, as this chapter has shown, these tensions were played out in the domain of popular culture and in the public sphere of the streets. Places of popular entertainment, such as the music halls, and public affronts to private virtue, such as prostitution, were issues of intense contemporary concern.

Similar issues emerge today over the control of popular culture in 'moral panics' around issues of football hooliganism, sexuality, racism, and riots (discussed, respectively, in Chs. 3, 4, and 6). Many of these issues involve a directly spatial dimension where the control of space is a crucial element in the maintenance of social order and the transformation of existing social relations. The domain of popular culture is a key area in which subordinate groups can contest their domination. It is, as Stuart Hall has argued, one of the sites where political struggle is engaged:

> It is partly where hegemony arises, and where it is secured. It is not a sphere where socialism, a socialist culture – already fully formed – might be simply 'expressed'. But it is one of the places where socialism may be constituted. That is why 'popular culture' matters. Otherwise, to tell you the truth, I don't give a damn about it (Hall 1981, p.239).

This may well be sufficient reason to be interested in popular culture. But the present chapter has taken a specifically geographical line on the spatial constitution of popular culture and its significance for the politics of class. But why, it may be asked, has this chapter focused on the 19th rather than the 20th century? It is not simply that the 19th century witnessed certain specific changes that came to define the popular in a particular way: the emergence of new classes, particular patterns of social segregation associated with the growth of cities, and the emergence of leisure as a separate category of social life for significant numbers of people – important though these changes were. More important is the insistence that *popular culture can only be approached historically*.

One final example illustrates the way an historical perspective can throw light on a contemporary issue. The current fascination with televison soap-operas like *EastEnders* can only be understood in terms of the actual disappearance of the 'communities' such programmes represent, just as the popularity of *Coronation Street* coincided with the physical disappearance of back-to-back housing in northern industrial cities like Salford and their eclipse by new suburban developments, represented in a younger generation of soap operas like *Brookside* or *Neighbours*.[10] Combining a geographical approach with an historical sensibility allows one to move beyond the level of contemporary representations of working-class life (Laing 1986) towards a more ethnographic understanding of its symbolic meaning. This is the direction in which future geographical work on popular culture may most profitably be orientated.

Notes

1 For an introduction to their work, see Held (1980, chapter 3).
2 In more positive versions of the popular, audiences are credited with an ability to transform and subvert the material of 'mass culture', implying an heroic as opposed to a pessimistic view of consumption.
3 Janowitz (1975) highlights the inconsistency with which the term is used, referring alternately to the imposition of control from above and to the process of internal self-regulation. Stedman Jones (1983), on the other hand, rejects the term as too static, implying that equilibrium is the norm from which conflict is a temporary aberration prior to the re-imposition of 'social control'. By contrast, Marxist analyses imply that conflict is pervasive and that society tends permanently towards a state of crisis. Stedman Jones concludes that 'social control/class expression' is a 'tempting but tautological couplet' (ibid. p.89).
4 On the development of the YMCA in the United States as an attempt to instil moral order among the disorganized urban masses, see Boyer (1978, Ch. 7).
5 Apart from biographical studies of individual performers, Bratton (1986) has edited a series of essays on 'performance and style'. For a general social history, Bailey (1978) remains the most informative as well as the most entertaining. The introduction to the paperback edition of his book, entitled 'Leisure, culture and the historian: confessions of a vulgar culturalist' (Bailey 1987), usefully updates the study and addresses some recent controversies in the field of cultural studies and social history.
6 Such categories served to mask the material connections between wealth and poverty in a fundamentally unequal society. Writing in 1845, therefore, Engels was able to turn the charge of depravity against the self-righteous middle classes of whom he declared: 'I have never seen as demoralized a social class as the English middle-classes. They are so degraded by selfishness and moral depravity as to be quite incapable of salvation. . . So long as they are making money it is a matter of complete indifference to the English middle classes if their workers eat or starve' (Engels 1958, pp.311–12).
7 The incident at Morant Bay became a *cause célèbre*. Governor Eyre brutally suppressed the rebels whose leader was summarily executed without trial. Victorian society divided in its reaction to the atrocity: Ruskin, Tennyson, Kingsley, Dickens, and Carlyle supported the Governor; John Stuart Mill, Darwin, Huxley, Spencer, and Lyell condemned him. The episode itself and domestic reactions to it can be pieced together from the accounts in Bolt (1971) and Jones (1980).
8 The New York-based Children's Aid Society proposed such a policy in the 1850s according to Stansell (1986, Ch. 10).
9 Stedman Jones provides a long list of these natural, environmental images applied to the urban poor including dens, swamps, deeps, wilds, abysses, shadows, netherworld and darkest regions (1974, p.463).
10 Other contemporary soap operas, such as *Dallas* and *Dynasty*, are harder to

understand in these terms. Chambers (1986) suggests that *Dallas* is 'kitsch' while *Dynasty* is 'camp', and it is tempting to dismiss them both as escapist fantasies. But the process of consumption is more complex, with audience reactions varying according to their social circumstances defined in class, gender, and national terms. Similarly complex reactions greeted British and American films in the 1930s, particularly in the North of England and Scotland, where 'the lah-di-dah accents of British actors aroused derision and hostility while the racy vivacity of American slang quickly became a vital part of popular culture' (Murphy 1983, p.98).

Chapter five
Gender and sexuality

Despite some recent welcome developments, gender and sexuality are still too rarely regarded as part of the central agenda of human geography. In cultural geography, they are even further from the mainstream, regarded as peripheral, private, and personal issues, not suitable for academic debate or public discussion. Geographers have found it convenient to hide behind the shield of a parochial definition of disciplinary boundaries. What could possibly be *geographical*, critics imply, about such intimate, personal subjects as gender and sexuality? This chapter addresses these questions and challenges the assumptions that lie behind them. It welcomes the general awakening of a social science interest in sex-gender systems and considers its implications for a reconstituted cultural geography. For changing ideas about sexuality and gender represent one of the most fundamental social changes of present times and one that involves cultural as well as political and economic dimensions. These changes have forced a recognition that the distinction between the private and the public is not fixed and immutable but culturally specific and socially constructed.

These introductory remarks help to establish an agenda for cultural geography that is much broader than that previously recognized. In particular, this chapter will argue that studies of gender and sexuality cannot take place in theoretical isolation but involve a radical transformation in conceptions of society as a whole[1]. The achievements of the women's movement and the political demands of gay liberation require more than a 'geography of women' or studies of isolated 'gay ghettos', important though these issues may be in themselves. For, as Stuart Hall has argued for cultural studies in general, feminism demands a major rethinking in every area of substantive work (Hall 1980a). Both gender and sexuality can be viewed as a mirror that reflects the broader structure of social relations. Their distinctive geographies provide one point of entry into a range of cultural questions of key contemporary significance.

Feminist approaches have influenced the practice of geography in a variety of ways (see the reviews by Lee & Loyd 1982, Zelinsky *et al.* 1982, Women & Geography Study Group 1984). Starting at the professional level, with the discipline of geography itself, a number of inquiries have been made into the status of 'women in geography', assessing their relative under-representation, particularly in more senior posts (cf. Zelinsky 1973b; McDowell 1979). Other studies have explored the 'geography of women' (e.g. Tivers 1978), particularly in the sense that urban planners and other bureaucrats have based their policies on a model of gender relations that implies and perpetuates female subordination (McDowell 1983). Feminist geographers have also considered the implications of the gender division of labour for understanding the differential impact of industrial restructuring in different localities, focusing on changing gender roles and relations (cf. Massey 1984, Lancaster Regionalism Group, 1985, Bowlby *et al.*, 1986, Foord *et al.*, 1986). Recently, too, geographers have begun to examine the geography of gender in the Third World and women's housing needs and experiences in cross-cultural perspective (Momsen & Townsend 1987, Moser & Peake 1987). Most recently, they have returned to the theme of gender and the urban environment (Little *et al.* 1988). The fundamental implications of the separation of 'home' and 'work' for the reproduction of patriarchal gender relations and the geographical ramifications of the domestic labour debate still remain to be fully explored.[2]

Although feminist geography in Britain has tended to focus on economic rather than on cultural issues, this chapter includes a discussion of the socially constructed world of 'women's work', raising a number of questions about the cultural definition of skill, the shifting boundaries of public and private space, and the intersection of patriarchy and capitalism in particular places at particular times. It begins, though, with a discussion of the significance of feminist theory for cultural geography.

Feminist theory and cultural geography

Contributions by geographers to feminist theory have been comparatively rare (though see Foord & Gregson 1986, and the subsequent debate in *Antipode*). The absence of such a theory, anchored in a critical examination of the nature of patriarchy, will certainly impair the development of a more adequate understanding of the whole gamut of situations in which unequal gender relations are structured and reproduced. For, as Kate Millett has argued, sexual dominion is underpinned and upheld by perhaps the most pervasive ideology of our culture, providing its most fundamental concept of power (Millett 1977, p.25). She goes on to argue that patriarchy as an institution is so deeply entrenched that it runs through all other political, social, and economic forms, whether of caste or class, feudalism or bureaucracy, just as it pervades every major religion.[3] It also exhibits great variety in history and locale.

That women's oppression by men takes a particular form in capitalist societies leads to the recognition that Marxism may offer a complementary rather than a contradictory agenda to that of feminism. But, as Michel Foucault's research on the history of sexuality reveals, there is no direct relationship between the rise of industrial capitalism and the control of sexuality (Foucault 1976). Rather, to quote Jeffrey Weeks: 'Capitalist social relations do certainly set limits and pressures on sexual relations as on everything else; but a history of capitalism is not a history of sexuality' (1985, p.6). Stuart Hall develops this argument in a most constructive way. Accepting that capitalism and patriarchy have 'distinct histories, different conditions of existence, different cross-cutting effects and consequences, which make impossible any neat alignment or correspondences between them', he nonetheless insists that 'a theory of culture which cannot account for patriarchal structures of dominance and oppression is, in the wake of feminism, a non-starter' (Hall 1980a, p.39). Understanding women's oppression clearly involves highly complex intersections between patriarchy and capitalism (Barrett 1980). Rather than simply reducing one to the other, however, capitalism and patriarchy are increasingly being theorized in terms of independent structures, with their own autonomous effects (e.g. Walby 1986).

Perhaps the most fundamental contribution of feminism to social theory has been the recognition that gender divisions (including so-called 'masculine' and 'feminine' personality traits) are *socially constructed*. This has been an historical process in which the socially constructed domain of gender relations has progressively won ground from the territory once occupied by genetically-based theories of sexual difference. Restraints formerly placed on women's actions (governing every form of behaviour from bicycling to voting) have increasingly been shown to have their roots in political and economic relations rather than in the laws of biology. Patriarchy's biological foundations are very insecure, while gender distinctions are overwhelmingly cultural (Millett 1977, pp.29-31). Differentiating between sex and gender therefore takes on crucial political significance.

While for some time it was convenient to assume a distinction between the socially constructed categories of gender and the biologically given categories of sex, even this distinction can no longer be sustained. Connell (1987), for example, challenges the assumption of sexual dimorphism in human beings, citing as evidence the results of chromosome testing of Olympic athletes at the Mexico Games in 1968. Transsexualism and sex-change operations also suggest that the biological categories of male and female, around which sexual and gender identities are built, are socially constructed.

Theorizing gender is therefore a highly contested domain of fundamental significance to feminist politics. For, as Connell (1985) has argued, the uncritical acceptance of a categorical distinction between women and men has directed feminist research away from other divisions (like class, race, nationality, and age) that cut across and complicate such deceptively simple oppositions. Other distinctions, such as those between sex and sexuality, and between gender and sexuality, also require careful theorization before being applied to particular geographical problems.

A feminist theory of cultural geography should begin by recognizing that sexual relations have a crucial and frequently neglected political dimension. One of the most lucid and sustained accounts of this position is Kate Millett's *Sexual politics* (1977), which provides a theory of sexual politics, an historical analysis of the 'sexual revolution', and a discussion of its literary reflection in the work of four representative authors. The literature of gay liberation has also produced forceful statements of the proposition that 'sex is politics', such as Gore Vidal's essay by that name. According to Vidal, sex is political not just in the sense that it may decide elections (as has arguably been the case with the Equal Rights Amendment, abortion, and gay rights in the United States), but also in the sense that sexual attitudes result from political decisions often mediated culturally by religious precept. In Vidal's own words:

> Any sexual or intellectual or recreational or political activity that might decrease the amount of coal mined, the number of pyramids built, the quantity of junk-food confected will be proscribed through laws that, in turn, are based on divine relations handed down by whatever god or gods happen to be in fashion at the moment (Vidal 1983, p. 190).

The first half of this chapter attempts to illustrate how feminist theory can be brought to bear on the study of cultural geography. Three case studies will be used to show some of the ways in which gender relations are both socially and spatially constituted. The first concerns one of the multiple ways in which women have been 'hidden from history', in this case concerning the exclusion of women's experiences from conventional accounts of American frontier history. The second concerns the interplay between capitalism and patriarchy in various forms of female employment in Britain. The third concerns a particular field of (largely female) employment, prostitution, which provides an example of the 'geography of women' and the contradictory nature of patriarchal oppression in a society that demands prostitution while simultaneously trying to repress it.

The second part of the chapter then develops some of the geographical issues that surround the question of male homosexuality. This is not to imply that the concept of gender applies only to women or that sexuality concerns only gay men. Neither is, in fact, the case. A cultural geography that is sensitive to questions of gender must recognize the extent to which *all social relations are gendered*, just as studies of sexuality must be sensitive to every aspect of human sexuality, recognizing that 'heterosexual', 'homosexual', and 'bisexual' are labels that grossly oversimplify a highly complex range of human behaviours and social practices. While much of the research on homosexuality has been by and about gay men (reflecting the gender relations of academia where men remain more powerful than women irrespective of sexuality), the contribution of lesbians both to the politics of gay liberation and to the academic study of sexuality should not be overlooked. The distinction between gender and sexuality is also rather blurred in practice, as patriarchal oppression involves the attempt by men to control women's sexuality through a host of practices (from the regulation of abortion, childbirth, and fertility, through the institutions of marriage, family, and private property, to more subtle strategies designed to enforce compulsory heterosexuality).

Having clarified some of the theoretical issues about gender and sexuality and explained the organization of the chapter, it is now possible to consider some examples, beginning with several that examine the cultural dimensions of economic exploitation and the significance of the gender division of labour for the oppression of women. This should not be taken to imply that patriarchy can be reduced to a system of economic exploitation or that women's oppression can be explained simply in terms of its functionality to capital. Indeed, as the following examples demonstrate, a reconstituted cultural geography must recognize the complexity of patriarchal oppression which operates through a range of economic, sexual and other practices.

The female frontier in America

In a pair of fascinating books, entitled *The lay of the land* (1975) and *The land before her* (1984), Annette Kolodny has embarked upon a revisionist history of American attitudes towards Nature through an analysis of the rôle of feminine metaphor in American life and letters, challenging the virtual absence of women from the mythology that surrounds the American frontier experience. Her work combines literary and cultural history informed by feminist theory and deep political commitment.

The original impulse for writing *The lay of the land* was of a particularly immediate political kind: the author's distress and indignation at her country's neglectful and exploitative attitude towards the environment as exemplified in the Battle for People's Park that was being waged in 1969 around the University of California's campus in Berkeley. Beginning from this point, Kolodny goes on to trace the roots of such conflicts in the metaphorical representation of the land as a feminine entity throughout American history. She documents the numerous and contradictory ways in which men imbued the land with the attributes of Virgin, Mother, and Mistress. By surveying a broad sweep of American literary history she shows how the New World was consistently envisaged as a source of male gratification, a virgin land to be tamed if not wilfully violated by the forces of male aggression. Even for 20th-century authors, like Norman Mailer in *Armies of the night* (1968), America is still represented as 'a beauty of magnificence unparalleled', 'a beauty with a leprous skin . . . heavy with child' and as a 'tormented lovely girl' – a series of images that reflects the range of conflicting feelings that Americans currently project on to the landscape.

Kolodny traces the web of images that centre on the metaphor of the land-as-woman, such as the series of oppositions in the 19th-century thought between passive filiation and active impregnation:

> For, just as the growing child must confront and mediate between his conflicting drives for individuation and maternal union, so, too, the American literary imagination found itself forced to choose between a landscape that at once promised total gratifications in return for passive and even filial responses and yet, also, apparently tempted, even invited, the more active responses of impregnation, alteration, and possession (Kolodny 1975, p.71).

Kolodny suggests perhaps too close a connection between the symbolic forms of literary imagery and the political values and social practices they reflect, but the overall pattern which she uncovers is remarkably consistent. The political lesson she draws from her literary and historical analysis is that people are still bound by the vocabulary of the feminine landscape. She suggests, however, that a sensitivity to such language provides at least the possibility of an alternative reading of the American past and of an alternative agenda for the future:

> It gives us . . . at least some indication of *how* those peculiar intersections of human psychology, historical accident, and New World geography combined to create the vocabulary for the experience of the land-as-woman. And it gives us, more importantly, another vantage point from which to understand those unacknowledged but mutually accepted patterns by which Americans have chosen to regulate their lives and interactions for over three hundred years now (ibid. pp.146–7).

Kolodny concludes that American historical experience and its literary expression have so far failed to provide a model either for a mature masculinity or for responding in a satisfactory way to the supposedly feminine qualities of nature.

The land before her takes up the themes of Kolodny's first book but develops them more subtly through a closer reading of a more limited literature. Rather than presenting a general argument about the role of metaphor in American literature and culture, Kolodny is here concerned to explore the way that American frontier imagery was 'shaped by personal psychology, social context and changing geography' (Kolodny 1984, p.xii). Apart from her revelation of the general absence of women's experience from the historical record of western expansion, Kolodny's detailed literary analysis reveals the existence of a pair of contrasting images that correspond to the differential male and female experience of the frontier. Masculine images stress the existence of a virgin land to be 'taken' and 'possessed', a Garden of Eden with infinite possibilities of exploitation; the feminine image, by contrast, is of a garden to be cultivated and domesticated, an intimate sphere centred on the family:

> After initial reluctance at finding themselves on the wooded frontiers of the northeast and the Ohio valley, women quite literally set about planning gardens in these wilderness places. Later, they eagerly embraced the open and rolling prairies of places like Illinois and Texas as a garden ready-made. Avoiding for a time male assertions of a rediscovered Eden, women claimed the frontiers as a potential sanctuary for an idealized domesticity. Massive exploitation and alteration of the continent do not seem to have been part of women's fantasies. They dreamed more modestly of locating a home and a familial human community within a cultivated garden (Kolodny 1984, p.xiii).

Kolodny charts the material struggle of women to make a home under the trying conditions of frontier life and the parallel struggle of the imagination to find an alternative language in which to express that experience. At least until the 19th century, American women seem to have been the unwilling inhabitants of a metaphorical landscape that they had no part in creating. Then, 'to escape the psychology of captivity, women set about making their own mark on the landscape, reserving to themselves the language of gardening' (ibid. pp.6-7). The vocabulary of confinement revealed in contemporary correspondence ('shut up with the children in log cabins') played a much larger part in women's imagination than the language of sexual conquest. Kolodny also explains the extraordinary popularity of books like Mary White Rowlandson's account of her capture by Narrangansett Indians, published in 1862, in these terms.

She shows how, at a later date, the domestic novels of western relocation by Maria Susanna Cummins, Emma Southgate, and Caroline Soule (published in the 1850s and '60s) served the ideological purpose of downplaying the differences between familiar New England landscapes and the wilder prairies of the relatively untamed West. While not so unashamedly promotional as the works of Mary Austin Holley or Eliza Farnham, published a few decades earlier, they nonetheless managed to trivialize topographical differences, hinting at the possibility of a speedy transition from log cabin to framed house or honeysuckled cottage, assuaging women's understandable fears of geographical isolation (ibid. pp.173-5).

Kolodny's work also has another relevance, exploring the extent to which women participate in constructing the gender ideologies by which they are oppressed, as well as the possibilities for resisting them. For women are not simply the passive recipients of male oppression. In subscribing to traditional gender roles of mother, wife, and daughter or to conventional ideologies of family, marriage, and home, women play an active part in reproducing the structures that oppress them. In the context of the American frontier, for example, the heroic status ascribed *by men and women* to actors like John Wayne in countless 'Western' movies can certainly be interpreted in this way. And, as Neil Smith has shown, frontier ideologies are extraordinarily persistent even in the contemporary city, where they reappear as ideologies of 'pioneering' or 'homesteading' in the urban 'wilderness' (Smith 1986).

Women's work?

A second example concerns the gender division of labour, its material consequences and the ideologies that sustain the notion of a separate sphere of 'women's work'. For the idea that some kinds of work are appropriate exclusively to men or to women is a peculiarly tenacious one for which little justification can be found in human biology. The 'heavier musculature of the male' referred to by Kate Millett (1977 p.27) explains neither their continued political supremacy nor their disproportionate share of high-status jobs. Female subjugation rarely depends on physical strength but relies instead on a system of ideological domination, upheld by a range of exclusionary and oppressive practices. There are very few occupations from which women are excluded on the basis of their physiological constitution and fewer still for which they are uniquely qualified by biology. Even the so-called 'maternal instinct', which is regularly employed to justify the confinement of women to the home and to restrict them to certain occupations for which their 'feminine' skills allegedly equip them, has no sound scientific basis. Men are no more suited biologically to careers in engineering and science, medicine, and law, or in Parliament for that matter, than women are uniquely qualified for the tasks of child-rearing, nursing, teaching, or secretarial work. Yet the differential representation of men and women in these professions confirms the existence of a systematic bias that works consistently in favour of men. Even within the same occupations, women frequently earn less than their male colleagues.

Historically, unskilled work has almost invariably been defined as feminine; if the same job is taken over by men it is redefined as skilled work. Cross-culturally, too, 'women's work' is consistently downgraded; the same job, performed by men in one society, is culturally and economically devalued when it is performed by women in another:

> Men may cook, or weave, or dress dolls or hunt hummingbirds, but if such activities are appropriate occupations of men, then the whole society, men and women alike, votes them as important. When the same occupations are performed by women, they are regarded as less important (Margaret Mead, quoted in Rosaldo & Lamphere 1984, p.xiii).

Cynthia Cockburn has explored the social definition of skill and its implications for the gender division of labour in a variety of contexts involving the introduction of new technology in the clothing industry, the mail-order business, and hospital X-ray departments (Cockburn 1986). In each case, women were found to be operating the new equipment but their conditions of work had not been unequivocally improved. They were under increased pressure to make the maximum use of expensive new equipment, they were vulnerable to further technological change, designed to be increasingly labour-intensive and they were excluded from higher status jobs such as maintenance technician and systems technologist. Nor were women to be found 'upstream' where the new technologies were being developed. Cockburn shows that women have not been excluded from these skilled occupations because the work is dirty, heavy, or dangerous. Rather, she suggests, more subtle, informal practices are involved. Men form friendship networks at work, based on the mutual exchange of knowledge and a kind of competitive humour that is implicitly masculine. Their career patterns assume the absence of family commitments and they carry over from the home a division of labour that defines technological competence as a masculine skill. Technological job segregation by sex clearly involves both economic and cultural dimensions. Men's higher earnings, social status, and skills are established not only at work but also as an extension of the pattern of gender relations determined in the home.

What, though, are the geographical effects of the gender-typing of occupations and skills? A provisional answer can be inferred from the work of Linda McDowell and Doreen Massey who discuss the ways in which patriarchy and capitalism interact to produce different patterns of gender relations in different places and at different times (McDowell & Massey 1984). From the four examples they provide, ranging from agricultural gang-work in East Anglia to cotton-spinning in Lancashire, only two examples will be discussed here.

In the first of their examples, McDowell & Massey focus on Britain's traditional coal-mining areas, characterized by an extreme sexual division of labour and by an archetypally 'masculine' industry. The exclusion of women from underground mine-work was guaranteed under the Mines (Regulation) Act of 1842, leading to the decline of the 'family labour' system. It was later extended by a form of union organization that was exclusively male. Women have therefore tended to be a highly marginalized part of the labour force and virtually absent from other aspects of public life. That these patterns are not immutable was, however, well illustrated during the miners' strike of 1984–5 when women played a crucial role through the network of Miners' Support Groups that grew up over the bitter months of the dispute. Women were unusually prominent in the strike (Massey & Wainwright 1984). Through it, many women got their first taste of political action and began to challenge not only the class subordination that they shared with men but also the double oppression that they experienced as working-class women (Loach 1984). The strike had a long-term effect on gender relations in these areas even if some households have since reverted to a traditional patriarchal division of labour.

The mining industry has also been the subject of an intense debate over the relative significance of class and gender in explaining women's oppression. Jane Humphries reviews the classic argument that women's oppression is functional to capital, reproducing labour power, creating additional surplus value, and ensuring political stability (Humphries 1977, 1981). She focuses on the family as an arena of production, on the material benefits it imparts in the process of class struggle, and on its function in the renewal of labour power. In contrast to her emphasis on the exclusionary tactics that men employ in the workplace, and hence on the adequacy of capitalism as an explanation of women's oppression, Jane Mark-Lawson and Anne Witz argue that men's position of authority at work derives, in no small part, from patriarchal relations constituted outside the workplace (Mark-Lawson & Witz 1986). They therefore question Humphries' insistence on the adequacy of a Marxist analysis of capitalist class relations for an understanding of women's oppression in the workplace and at home.

The second example from McDowell & Massey's work concerns the system of sweated labour in the rag-trade of London's East End. Here, the spatial structure of patriarchal households actively supports capitalist relations of production by discouraging unionization among the geographically dispersed labour force. Much of the garment trade relies on poorly paid home-work, a low status form of employment that has traditionally been occupied by immigrant women (first Jewish and now Asian). Patriarchy and capitalism combine in a mutually reinforcing way, so that working-class men do not feel their status as 'bread-winner' is compromised since their wives work at home, while the dispersal of production that this implies helps employers to keep wages down by inhibiting organized resistance to exploitative wage rates.

In her analysis of the geography of production, Massey (1984) adds a further example of her own (as well as providing a more extended discussion of the coalfields example). Focusing specifically on the interaction of class and gender, she examines the changing employment structure of South-West England where the traditional pattern of small industries and self-employment in the holiday trade, with low female activity rates in industry, has gradually given way to a branch-plant manufacturing economy, with a 60 per cent female labour force. The traditional employment structure offered women few opportunities for paid work; employment was often seasonal and unemployment was generally high. Moreover, the dispersed population in rural areas gave further support to the structure of patriarchal gender relations, with female activity rates in urban areas some nine per cent higher than the rural average.

During the 1960s, however, the existence of a large female labour reserve began to attract new industry, particularly in light manufacturing (instruments and electrical engineering) and in services. Local capital (in the guise of the Cornwall Industrial Development Association) opposed the arrival of new investors who threatened to disrupt the traditional economy. In making their case, the Association argued that the creation of more factory jobs for women went against 'the natural preference of the people'. They supported instead the creation of more jobs for women *in the home*. An argument over the spatial form of the labour market (increasingly urban and factory-based as opposed to rural and home-based) was articulated through the ideology of the family (and people's 'natural' preferences) with specific effects on the sexual division of labour and on gender relations in general.

These examples illustrate the complex ways in which patriarchy and capitalism intersect to create historically and geographically specific patterns of class and gender oppression. They show that class relations have a cultural as well as an economic dimension and that patriarchy cannot be confined to questions of sexuality, marriage, or domesticity. 'Home' and 'work' cannot readily be separated, as relations of dominance and subordination established in one domain carry over into the other. The following section considers a particularly complex area of 'women's work', female prostitution, where class and gender relations are clearly constituted geographically.

The landscape of female prostitution

It is no coincidence, as Richard Symanski points out in his book on female prostitution in Western societies, *The immoral landscape* (1981), that men pay more for illicit sexual access than for virtually any other form of female labour. His book provides one of the few exceptions to the general silence among geographers concerning such 'disreputable' areas of sexual relations.

Symanski shows that prostitution, like other forms of 'criminal' behaviour, closely reflects the structure of social relations in general (cf. S. Smith 1984b, 1986). He shows, for example, that prostitution has a hierarchical structure that corresponds to the class structure of the wider society. 'High-class' prostitutes (or 'call girls') visit their clients' home, hotel, or party; 'lower class' prostitutes (or 'street-walkers') work in more public places and take their clients to their own home, hotel, or rented room. Street-walkers are therefore the most vulnerable group and make up the majority of arrests. They are harassed by the police, not just because they are low status law-breakers but because their relatively high profile represents a visible affront to public morality (see also Ch. 4). Significantly, very few men are arrested for their part in the business of female prostitution because only the female side of the bargain is generally defined as illegal and actively enforced. Though some 50 per cent of street-walkers in the United States have pimps (almost all of whom are male), the latter form only a tiny fraction of those arrested in connection with prostitution while taking upwards of 90 per cent of the prostitutes' earnings (ibid. p.150). Male supremacy could scarcely be more blatant. Moreover, in a further parallel of the wider structure of social relations in America, black women are over-represented among low status prostitutes and relatively under-represented among higher status 'call girls', taking a disproportionate share of the risks that accompany the higher visibility of 'street-walking' (ibid. p.91).

The control of prostitution mirrors the regulation of sexuality in general which is often quite explicitly spatial. As Foucault shows in his history of sexuality, for example, Victorian society literally 'made room' for illegitimate sexuality by designating certain places, like brothels and mental hospitals, for certain specific practices:

> Only in those places would untrammelled sex have a right to (safely insularized) forms of reality, and only to clandestine, circumscribed, and coded types of discourse. Everywhere else, modern puritanism imposed its triple edict of taboo, nonexistence, and silence (Foucault 1976, pp.4-5).

In this case, a social rule was implemented through the combination of a particular spatial form and a certain mode of discourse. In the specific case of prostitution, the regulation of sexuality may be even more blatantly spatial. Repression and containment are the most usual 'geopolitical' strategies for the control of prostitution described by Symanski (1981). There is also a familiar geographical ring to his discussion of police tactics in Stepney during the 1940s when the suppression of prostitution in that area led to its displacement elsewhere (see Fig. 5.1). The dispersal of prostitution to more 'suburban' locations made it more difficult to police but the strategy was successful to the extent that it made it much less publicly visible. Similar processes have been described elsewhere. According to Shumsky and Springer (1981), for example, the attempted reform of prostitution in 19th-century San Francisco led only to a displacement of the 'red light' district rather than to a reduction of prostitution in general.

Symanski demonstrates that prostitution is a highly urban phenomenon with a distinctive geography at a variety of levels. In California, for example, the number of prostitutes (as measured by average yearly arrests) is directly related to the relative position of different places within the urban hierarchy. Symanski distinguishes a 'local and trucker

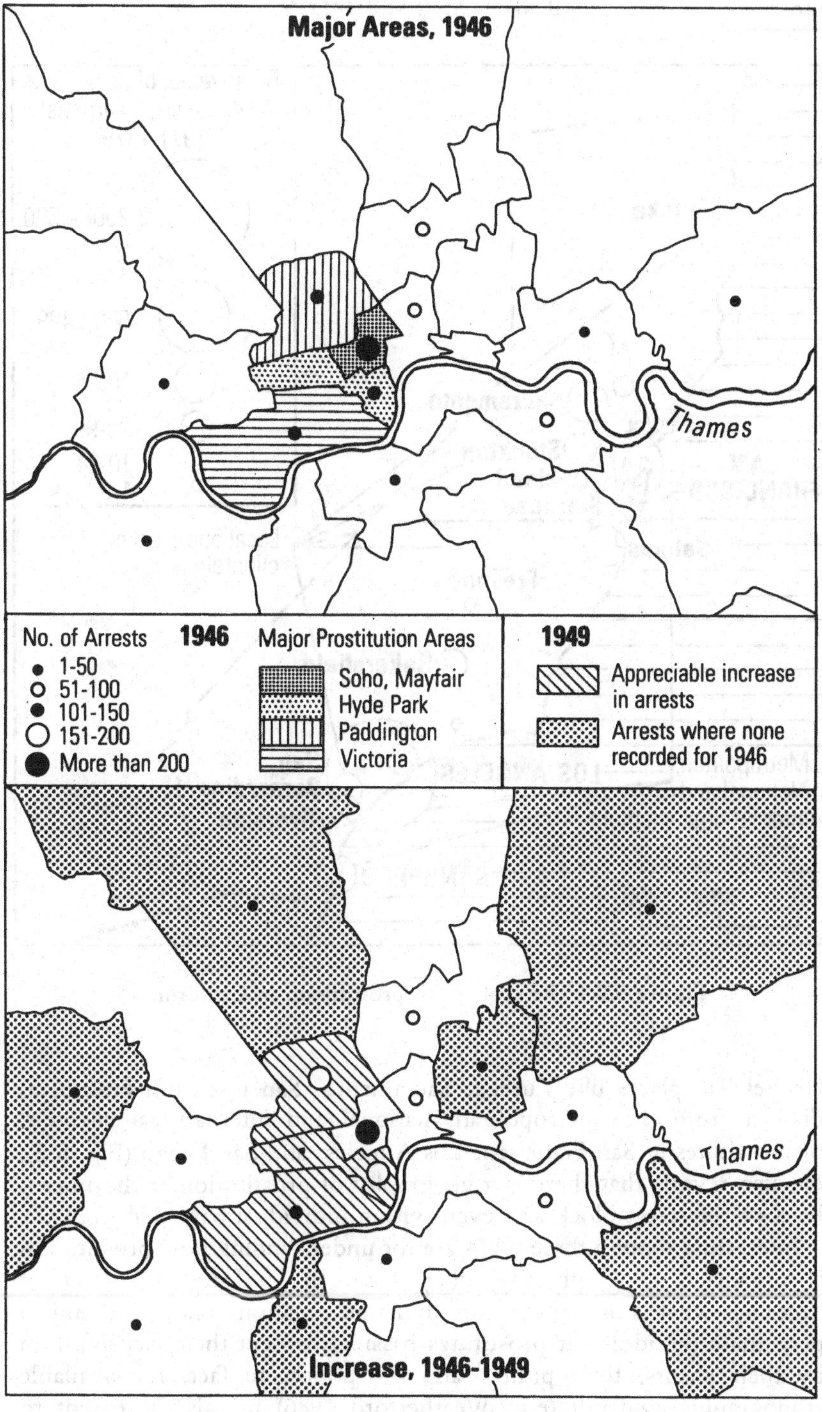

Figure 5.1 The displacement of prostitution in London, 1946–49

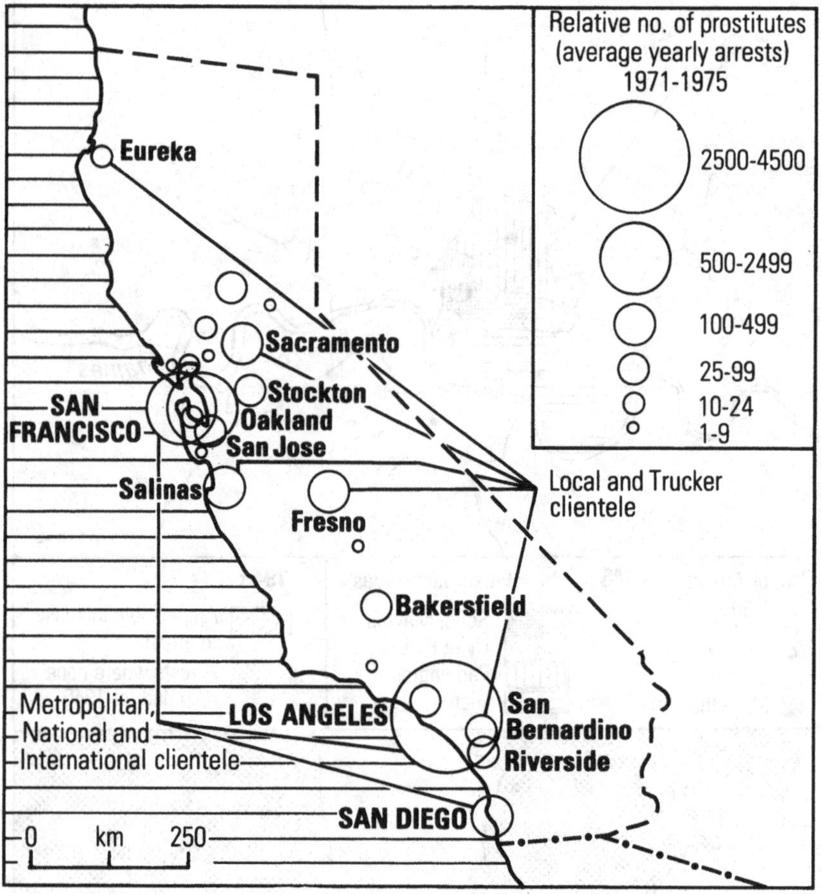

Figure 5.2 The ecology of prostitution in California

clientele' in places like Eureka, Sacramento, San Jose, and Fresno, as distinct from the 'metropolitan, national and international clientele' of prostitutes in San Francisco, Los Angeles, and San Diego (Fig. 5.2). He also reveals that there is a geography of prostitution at the micro-level of the street block and even within individual bars (Fig. 5.3).

Interesting though these maps are for understanding how prostitution is organized geographically, they have the unfortunate effect of 'freezing' what is in fact a highly dynamic situation. They also tend to perpetuate the idea that prostitutes passively accept their victimization by their clients, their pimps, and the police. In fact, the available ethnographic evidence (e.g. Weatherford 1986) reveals the extent to which prostitutes themselves adapt their spatial behaviour to prevailing circumstances, constantly moving between and within cities to

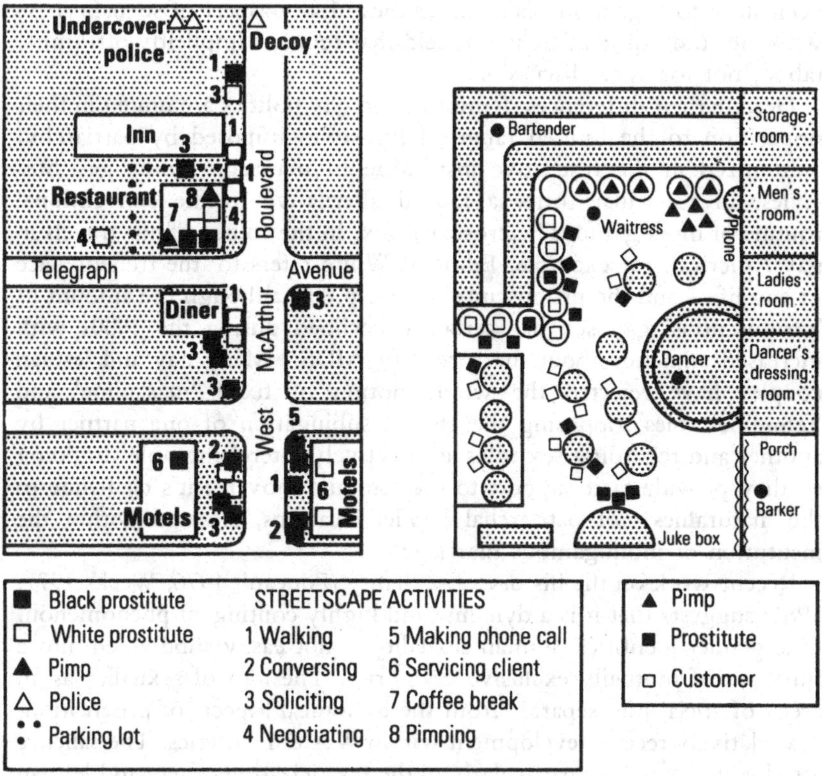

Figure 5.3 The micro-geography of prostitution

maximize their contact with potential clients, to minimize conflict with their pimps, and to avoid detection by the police. Though the police attempt to impose a spatial strategy of containment or dispersal, the prostitutes adopt mobility as a strategy to frustrate the process of law enforcement. If their pimps make intolerable demands or become excessively violent, they respond by moving to more favourable conditions elsewhere. There is a conscious and unconscious use of space by all concerned, even within the context of such a heavily asymmetrical power structure.

While feminism requires a complete reorientation of cultural studies to challenge the multiple ways in which women are subordinated to men in contemporary society, so in turn has the movement for gay liberation begun to challenge basic notions of sexual identity. In shifting the boundaries between private and public domains, politicizing the 'private' world of people's sexual relations, gay and lesbian politics are simultaneously social and spatial. To date, however, geographers have been slow to engage in academic studies of any aspect of sexuality. As with the study of gender, it is a field that remains implicitly off-limits, taboo, not for open discussion.[4]

Despite their differences, feminism and gay politics are united in their opposition to the limited range of life-styles imposed by patriarchy, epitomized in the restrictive institution of marriage. Some gay life-styles, in particular, offer a radical alternative to the conventional routine of monogamous heterosexual sex. In the course of his 'travels in gay America', for example, Edmund White refers to 'the frenetic pace of gay life – and the promiscuity' (1986, p.80), although the amount of 'promiscuous' gay sex has certainly declined during the 1980s with increased anxieties about the spread of AIDS.[5] Many gay and lesbian couples have rejected the whole notion of fixed 'masculine' and 'feminine' roles, opposing the eternal subjugation of one partner by another and regarding sex roles as effectively interchangeable. In doing so they provide tacit support to the feminist movement's challenge to the 'naturalness' of patriarchal gender relations, situated within the institution of monogamous marriage.

Recent work on the history of sexuality (Foucault 1976, Weeks 1977, 1981) suggests that it is a dynamic and highly contingent phenomenon. Like gender identities, human sexuality is not easily subdivided into a finite set of mutually exclusive categories. The idea of sexuality as an arena of social life, separate from the biological aspects of procreation, is a relatively recent development within Western societies. The salience attached to sexuality, particularly in the politicization of gay and lesbian identities, where sexuality is the prime basis of self-definition, is an even more recent phenomenon. It also has a distinctive geography, described in the following sections on the emergence of the 'gay ghetto' in San Francisco and other American cities.

The spatial basis of gay identity

Geographical studies of homosexuality have mostly been confined to questions of social segregation and to debates about the existence of various 'gay ghettos'. Several sociologists have also explored this idea. Martin Levine, for example, provides some very sketchy maps of the principal gay areas in a number of US cities, based on the entries in a national directory of gay gathering places (Fig. 5.4). Such areas he defines in terms of the concentration of gay institutions, a conspicuous and locally dominant gay subculture that is socially isolated from the larger community, and by a residential population that is substantially gay (Levine 1979, p.185). Laud Humphreys adds the necessity for a marked tolerance of homosexuality to this list of attributes that characterize the definitive 'gay ghetto' (Humphreys 1972, pp.80-1). From a geographical point of view, Barbara Weightman distinguishes a range of 'gay spaces' within the gay community, including gay bars, gay regions, gay resorts, and gay neighbourhoods, while introducing her own concepts of gay action- and activity-spaces (Weightman 1981, pp.107-11). Bonnie Loyd, for her part, dwells on the implications of the ghetto concept by drawing comparisons between the 'spatial strategies' of gay men and other 'minority groups' (Loyd 1982).

The geographical study of sexuality has only recently progressed beyond these rather unsophisticated contributions. As with other 'minority groups', the gay and lesbian communities have their own distinctive universe of discourse which provides a means of entrée to their social world and to its spatial constitution. The significance which politically active gay men and lesbians attach to *coming out* (from the safety, anonymity, but stultifying confinement, of the 'closet') provides just one example. Many gay men, in particular, have found it easier to make the symbolic public statement of their private sexual preference in the context of a gay neighbourhood which offers both practical and moral support. Paradoxically, however, gays may be perceived as more of a threat to the well-being of the heterosexual majority in places where they are most highly concentrated and socially visible. In other areas, they can be safely ignored as a socially insignificant and invisible minority. Much of the recent controversy surrounding Clause 28 of the Local Government Bill which forbids the 'promotion' of homosexuality concerns the toleration of gay sex *in private*, but the repression of its *public expression*. The 'liberal' attitude of condoning private homosexuality while condemning its expression in public is a key area of debate in the politicization of homosexuality and one which is defined above all in spatial terms.

Gay neighbourhoods therefore represent a powerful symbolic statement as well as a potentially effective electoral base. As Lauria and Knopp suggest, the social construction of gay identity in the US can be interpreted as a distinctively spatial response to an historically specific form of homophobic repression:

> Gays have done more with space than simply use it as a base for political power. They continually transform and use it in such a way as to reflect cultural values and serve the special needs of individual gays *vis-à-vis* society at large (Lauria & Knopp 1985, p.159).

At a different scale, gay men frequently refer to San Francisco as 'our town'. Edmund White, for example, uses this phrase as the title for his

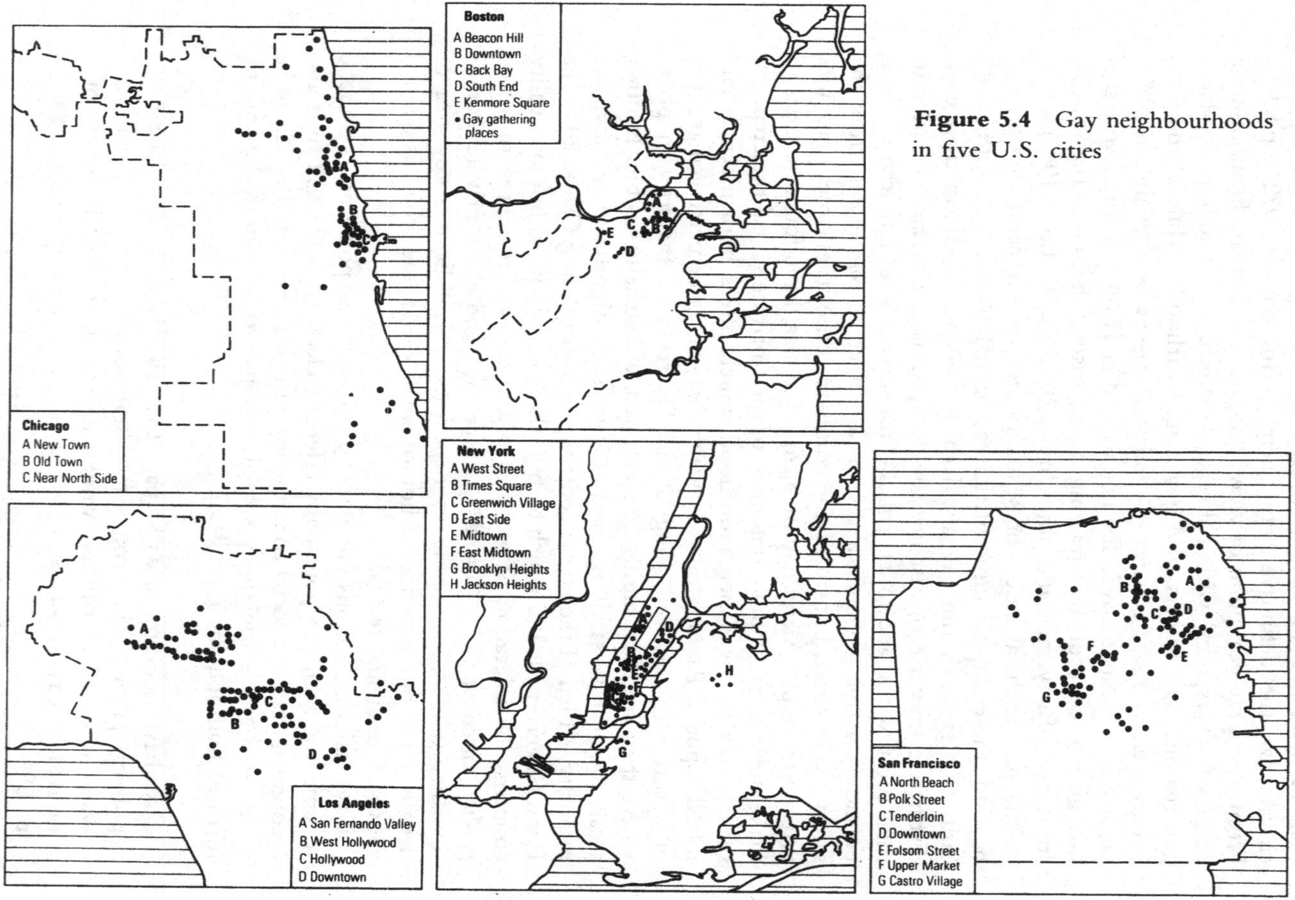
Chicago
A New Town
B Old Town
C Near North Side
Boston
A Beacon Hill
B Downtown
C Back Bay
D South End
E Kenmore Square
• Gay gathering places
Los Angeles
A San Fernando Valley
B West Hollywood
C Hollywood
D Downtown
New York
A West Street
B Times Square
C Greenwich Village
D East Side
E Midtown
F East Midtown
G Brooklyn Heights
H Jackson Heights
San Francisco
A North Beach
B Polk Street
C Tenderloin
D Downtown
E Folsom Street
F Upper Market
G Castro Village

Figure 5.4 Gay neighbourhoods in five U.S. cities

chapter on San Francisco in *States of desire* (1986), his autobiographical account of 'travels in gay America'. One of the reasons that San Francisco occupies this special place for gay Americans is that it was the first city in the US to elect an openly gay man to public office. It has since become a Mecca for gay men and women from throughout the world while simultaneously attracting a disproportionate amount of research into every aspect of homosexuality. San Francisco therefore provides the most readily available evidence on which to base an understanding of the spatial expression of sexuality and for gauging the significance of territory in the development of gay politics.

Gay politics and the restructuring of San Francisco

Manuel Castells includes a chapter on the relationship between cultural identity, sexual liberation, and urban structure in his monumental study of urban social movements, *The city and the grassroots* (1983). In other chapters Castells describes the importance of gender in urban political protests such as the Glasgow Rent Strike, or the combination of race and class in the American Civil Rights Movement of the 1960s. In the chapter on San Francisco, however, he is primarily concerned with the role of gay liberation in restructuring the city. Castells argues that sexuality cannot be considered in isolation from the wider matrix of ethnicity, class, and politics that forms the basis of San Francisco's social geography. The inclusion of sexuality as a key dimension in understanding urban change is, however, a relatively recent feature in both bourgeois and Marxist social science. Though some of Castells' critics regard his later work as a 'defection' from the position he outlined in *The urban question* (1977), Castells' own comments on the transformation in his work are more instructive:

> . . . although class relationships and class struggle are fundamental in understanding the process of urban conflict, they are by no means the only or even the primary source of urban social change. Our theory must recognise other sources of urban social change: the autonomous role of the state, gender relationships, ethnic and national movements, and movements that specifically define themselves as citizen movements (Castells 1983, p.xviii).

In turning to his analysis of the San Francisco experience it is useful to bear these recent changes in Castells' intellectual orientation in mind.

In 1980 San Francisco had an estimated population of some 115000 lesbians and gay men, comprising 17 per cent of the city's population and an even higher proportion (some 25 per cent) of registered voters. These figures are, of course, only speculative as Census data do not inquire about sexuality. Nonetheless, as Castells argues, gay San Francisco has become 'a powerful, though complex, independent community at spatial, economic, cultural, and political levels' (ibid. p.138). It is a city to which gay people come to learn how to be gay, to sample the range of life-styles that are available there in conditions of relative freedom and toleration. Just how San Francisco achieved this position in a national context that is still fiercely homophobic needs to be explained.

San Francisco has been described as an 'instant city', a settlement for adventurers attracted by the Gold Rush of 1849:

> San Francisco was always a place where people could indulge in personal fantasies and a place of easy moral standards. The city's waterfront and Barbary Coast were a meeting point for sailors, travellers, transients, and lonely people – a milieu of casual encounters and few social rules where the borderline between normal and abnormal was blurred. San Francisco was a gateway city on the western limits of the Western world, and in the marginal zones of a marginal city homosexuality flourished (ibid. p.140).

San Francisco became a focal point for the 'Beat Generation', celebrated in Jack Kerouac's ode to mobility, *On the road* (1958). It was a centre for the hippie subculture and the associated drug scene in the 1960s. Throughout this time the city was also developing an explicitly gay identity with a visible spatial expression in such neighbourhoods as the Castro. Castells maintains, however, that lesbians, unlike gay men, do not concentrate in a given territory but maintain their social identity through informal interpersonal networks (ibid. p.140). Though there is certainly some truth in the idea of the 'invisible lesbian' (a phrase employed by *Gay News* to draw attention to their lower purchasing power, their fewer businesses, and their lower level of politicization than gay men), there is also growing evidence of spatial concentration among lesbians as well as among gay men (see, for example, recent work by Larry Knopp (1987) on Minneapolis).

In looking at the way in which gay life-styles flourished in San Francisco it is important to trace the political, legal, and economic context (D'Emilio 1983) rather than see it as a purely 'cultural' phenomenon. An early advance in the legal status of gay people was won in 1951 when the California Supreme Court declared it illegal to close down a bar simply because it attracted a gay clientele. As a result, the number of gay bars in San Francisco shot up from 58 in 1969 to 234 in 1980. However, the legalization of homosexuality in California did not put an end to public intolerance of gay people and their meeting-places. Gay bars and bath-houses have traditionally been targets of harassment, including police raids such as those that provoked the Stonewall riots in Greenwich Village in New York in 1969. Repressive legislation continues to be introduced, such as the infamous Proposition 6 in California, in 1978, which sought to deny the right of gay people to teach in the state's public school system. While this particular Bill was eventually defeated, debates at the time revealed the depth of public ignorance about homosexuality and people's persistent intolerance towards gay people (expressed here in terms of parents' fears of child molestation, which happens to be an overwhelmingly heterosexual crime).[6]

Castells traces the evolution of a consolidated gay community in San Francisco (Fig. 5.5) which ultimately provided a sufficiently compact and tightly organized power base from which to launch a successful political campaign. While a number of gay candidates had previously run unsuccessfully for election to San Francisco's Board of Supervisors, the change of electoral system from at-large to district-based elections in 1977 provided the territorial platform necessary for the election of the first gay supervisor, Harvey Milk, representing the Castro (District 5). Harvey Milk ran a camera store in the Castro and had been instrumental in building up gay businesses throughout San Francisco by urging gays to 'buy gay'. With the support of the city's liberal mayor, George Moscone, the Board of Supervisors passed a far-reaching Gay Rights Ordinance in April 1978. Coupled with the defeat of Proposition 6 that same year, Harvey Milk was at the height of his political power and gay rights were similarly at an unprecedented high.

All this was brought to an abrupt and violent end in November 1978, when George Moscone and Harvey Milk were assassinated in their offices at city hall by fellow supervisor Dan White who had campaigned against gay rights and other progressive measures. Their deaths were immediately followed by a peaceful candlelit procession involving 20000 people in a march on city hall. A totally different reaction followed Dan White's trial, when he was convicted on two counts of 'voluntary manslaughter' but only given the minimum sentence of seven years in jail. The gay community responded by staging a violent protest outside city hall, setting fire to police cars, smashing windows, and causing several thousand dollars' worth of damage. In the months following Harvey Milk's brutal killing, the new mayor, Diane Feinstein appointed another gay politician, Harry Britt, as Supervisor for District 5. Dan White's prison term was eventually remitted for good behaviour but he later committed suicide after he was released from jail.

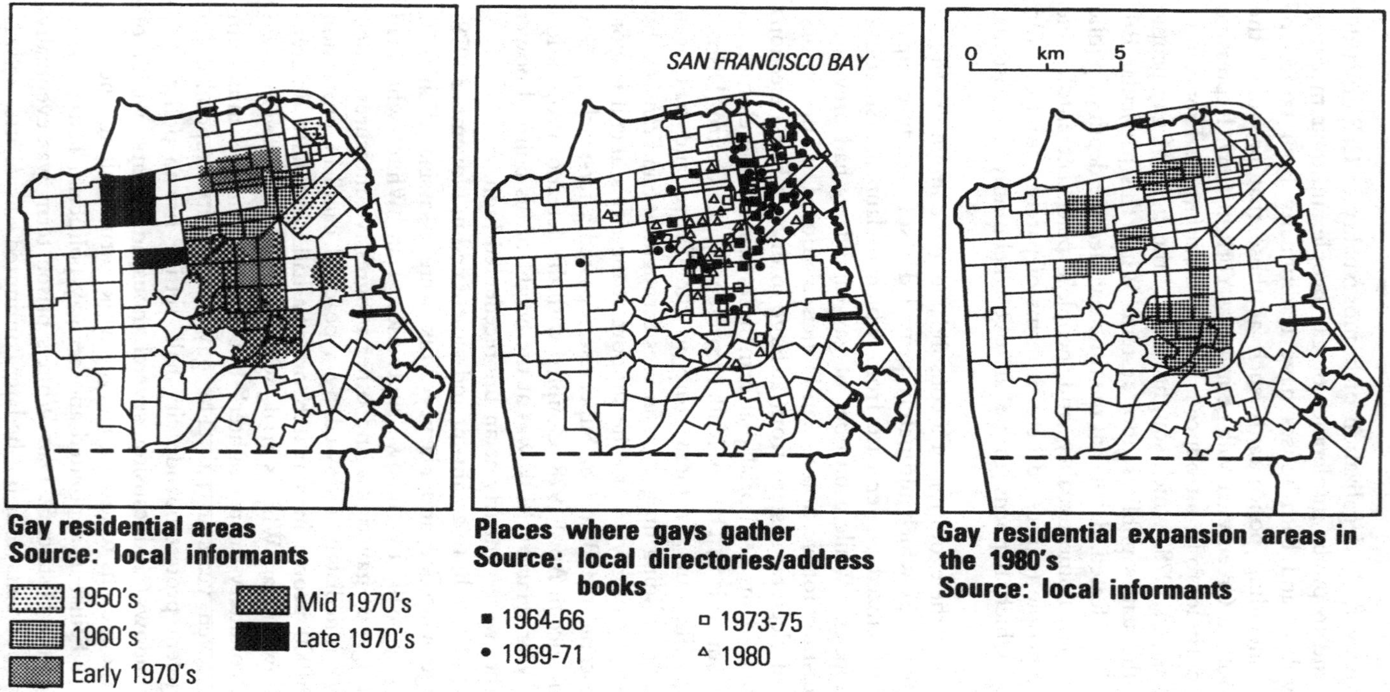

Figure 5.5 The development of gay social space in San Francisco

Before drawing any implications from these dramatic events (which subsequently became the subject of an award-winning documentary film, *The Times of Harvey Milk*, 1984), it is useful to consider the general context in which the development of San Francisco's gay community took place. The city originally grew up around the entrance to the natural harbour of San Francisco Bay in the mid-19th century and developed a highly specialized entrepôt function symbolized by the naming of the Golden Gate. Financial dealings have always played a major part in San Francisco's mercantile and commercial history, laying the foundation for the development of today's classic post-industrial economy, with a doubling of office space between 1960 and 1980. The city has become the headquarters for the finance, insurance, and real estate industries, particularly those that are orientated towards the economically booming Pacific Ring. At the same time, San Francisco's population has become truly international: 22 per cent Asian, 13 per cent black and 12-15 per cent Latino in 1980.

One result of this general process of social and economic restructuring has been the evolution of a so-called 'pink economy', taking its name from the pink triangle which Hitler forced homosexuals to wear in Nazi concentration camps (Heger 1980). Gay people employ each other for a variety of services from plumbing to furniture removals, as well as patronizing gay stores, businesses, and bath-houses. The Golden Gate (gay) Business Association had some 250 members in 1980, for example. Though there is a community of gay professionals (including lawyers, doctors, and therapists), the very attractiveness of the city to gay people leads to its own problems. The shortage of good jobs causes gays to be underemployed or over-qualified for the jobs they hold.

A further consequence of a city with more singles, more divorcees, and more gays than almost any other, is the creation of extreme pressure in the housing market. This has brought the gay community into conflict with lower income minority groups such as the Latinos. While gay men have been praised as 'the worker ants of our reviving cities', deserving credit for 'having made our inner cities safe and attractive centres once again' (White 1986, p.63), they have also been responsible for pushing black and Latino families out of their homes in areas like Haight-Ashbury and the Mission District where the renovation of housing by gay people has had the effect of pushing up rents beyond the means of poorer minority tenants. The role of gays in the so-called 'urban renaissance', and the displacement of working-class communities that followed, are problems that geographers have only recently begun to consider and that have yet to receive a satisfactory political resolution (Lauria & Knopp 1985)

The restructuring of gay San Francisco occurred through the combination of a sexual orientation, a cultural revolution, and a political movement with a firm territorial basis (Castells 1983). The inscription of a gay identity on the urban landscape is now an incontrovertible fact, despite the changes that have occurred as a result of the AIDS epidemic. Yet these achievements are as vulnerable as the gains of the 'sexual revolution' itself. The fragility of gay rights was revealed during the Reagan administration by the influence of the so-called 'moral majority'. Adverse publicity surrounding the AIDS epidemic, notoriously described by the popular press as a 'gay plague', has also set back the advancement of gay rights by several years and further conservative reaction can certainly be anticipated.

Finally, in considering the development of a spatially and socially conspicuous 'gay community' in San Franciso, it is worth remembering that only a fraction of the gay and lesbian population is represented within that particular community. If there are some 25 million gay men and lesbians in the United States (a conservative estimate based on Kinsey's famous ten per cent statistic), then those involved in urban reinvestment and other aspects of upward social mobility can only be a small proportion of the total gay and lesbian population. What usually passes for the 'gay community' is actually a minority of the minority – its most politicized and vocal fraction. Applying the same arithmetic, the gay and lesbian population of the United States also includes about 3½ million people below the federally-defined poverty line; four million malnourished people, many of them children; and 400 000 homeless people (Fair 1987). These are sobering statistics that demand to be taken seriously by the gay community if it is committed to repairing the range of social inequalities that characterize contemporary society, not just those that are defined in terms of sexuality.

Conclusion

The preceding discussions of gender and sexuality both contain important lessons for a revitalized cultural geography. Feminist theory issues a radical challenge to established distinctions between the private and the public, while gay liberation demonstrates the political nature of ostensibly 'cultural' movements. The development of gay politics in San Francisco provides particularly clear evidence of the territorial basis of urban social movements, while the analysis of gender relations in different geographical contexts demonstrates the way capitalism and patriarchy intersect in complex ways through space and time. But perhaps the most important lesson to be drawn from the study of gender and sexuality is that neither phenomenon can be understood in isolation from the wider social context. For gender relations are embedded in a matrix of social relations involving *both men and women*, while the study of sexuality cannot be confined to the study of gay men. Gender relations apply as much to men as to women, while the concept of sexuality applies as much to lesbians as to gay men, and no less to heterosexuals than to homosexuals.

The division of this chapter into separate sections dealing with gender relations (such as those that define a world of 'women's work') and sexuality (particularly concerning male homosexuality) should not be allowed to obscure the important links that exist between gender and sexuality. For example, just as there is an economic basis to the subordination of women (described in the foregoing examples concerning the gender division of labour and the world of 'women's work'), so there is an economic dimension to the role of gay men in the restructuring of San Francisco (revealed most clearly in their involvement in the gentrification process). While both gender and sexuality have important economic dimensions, neither can be understood in purely economic terms. They must also be understood as *cultural constructions* both in the way they are constituted and in the way they are subjectively experienced.

The remainder of this conclusion will attempt to draw out some of the more general points that can be made from the previous examples. For instance, the isolation of sexuality from other social issues in San Francisco is of political as well as theoretical significance. The development of a 'gay ghetto' in the Castro provided the power base for Harvey Milk's successful election, but the extension of the ideology of gay liberation to the rest of the city has been far more problematic. Castells is highly critical of the gay movement in San Francisco for its willingness to engage in traditional coalition politics, simply adding another strand to the local political system, thereby increasing 'the poverty of pluralism' (Castells 1983).

The distinction that Castells draws between the use of space by lesbians and gay men also merits further consideration. While Harvey Milk's successor, Harry Britt, has argued that 'When gays are scattered, they are not gay, because they are invisible' (Castells 1983, p. 138), gay politics must surely seek a wider arena than the 'ghetto' if they are to achieve a wider transformation of traditional power relations throughout the city. 'Gay ghettos' like the Castro can too easily be accommodated within traditional power structures, becoming just another attraction, like Chinatown or Fisherman's Wharf, on the itinerary of the San Francisco tourist. Worse still, territorial concentration also leaves gay people vulnerable to police harassment and other forms of repression. But the apparent reluctance of lesbians to form identifiable residential clusters has a variety of other explanations that further emphasize the overlap between gender and sexuality as bases of women's oppression. For lesbians are no less subject to patriarchal forms of exploitation than are heterosexual women, including their restricted access to private property. The difference may also reflect on the narrower political agenda of gay liberation, concerned mainly if not exclusively with homosexual rights, in comparison with the broader feminist agenda, concerned with every aspect of the relations between men and women, not just those that are defined in terms of sexuality.

Despite these differences, gay liberation and the women's movement have much in common. Both advocate the transformation of social life in general; neither is exclusively a 'minority' issue. The 'geography of women' and studies of the 'gay ghetto' may be welcome as a step in the direction of progressive social change. But neither is an adequate response to the range of questions prompted by the study of gender and sexuality. As this chapter has tried to show, a key area in the study of gender and sexuality concerns the redefinition of public and private space. A similar argument has been made with respect to popular culture in Chapter 4. It applies also to the study of race, the subject of the next chapter.

Notes

1 Geographers have much to learn here from historians and sociologists. See, for example, the first volume of the history of sexuality by Michel Foucault (1976), the sociological histories of Jeffrey Weeks (1977, 1981, 1985), and the History Workshop's collection of essays on *Sex and class in women's history* (Newton *et al.* 1983).

2 Among recent works that have addressed these issues, those by Pahl (1984, 1988) and Redclift & Mingione (1985) are probably most sensitive to the significance of space.

3 Other writers would challenge the primacy that Millett gives to patriarchy. Linda McDowell (1986), for example, argues strongly for a class-based explanation of women's subordination, questioning the theoretical separation of capitalism and patriarchy. Among recent attempts to situate the analysis of gender and sexuality in terms of contemporary social theory, Connell's *Gender and power* (1987) is outstanding.

4 This reticence among academic geographers to confront issues of sexuality may reflect their awareness of the possible adverse consequences of an explicitly spatial analysis of gay and lesbian communities. Mapping the addresses of gay and lesbian households or businesses, for example, might increase the dangers of police harassment. The controversy that surrounded Laud Humphreys' *Tearoom trade* (1970), a study of homosexual sex in public places, where the author used police records to locate the home address of those involved by tracing the registration numbers of their cars, is indicative of the extreme sensitivity of the issue.

5 Both Martin Amis (1986) and Frances FitzGerald (1987) provide some preliminary information on the impact of the AIDS crisis on gay life-styles. While generally positive, however, some of Amis' comments are less than helpful. He writes, for example, that: 'Gay men routinely achieve feats of promiscuity that the most fanatical womaniser could only whistle at' (Amis 1986, p.192). For a more constructive discussion of the AIDS crisis, see Altman (1986).

6 In Britain, the repression of homosexuality has similarly waxed and waned, from the generally liberal climate of the 1960s to the more repressive 1980s. The changing fortunes of gay liberation in Britain can be traced in the sarcastic lyrics of Tom Robinson's song, *Glad to be gay*, originally recorded in 1978 and periodically updated.

Chapter six
Languages of racism

> Can you divide human reality, as indeed human reality seems to be genuinely divided, into clearly different cultures, histories, traditions, societies, even races, and survive the consequences humanly?
>
> (Edward Said 1978)

The social construction of 'race'

Although racism often has intensely practical consequences, ranging from discrimination to verbal abuse and physical violence, these practices are informed by racist ideologies which are themselves worthy of further investigation. One point of entry into this intangible realm of ideas and beliefs is through the study of the language by means of which racist ideologies are routinely expressed (see also, Ch. 7). In this chapter, a broad definition of language will be employed, including visual as well as verbal representations. The approach is historical, for racism is not a uniform or invariable condition of human nature but, like other ideologies, is firmly rooted in the changing material conditions of society. As Sivanandan has argued:

> Racism does not stay still; it changes shape, size, contours, purpose, function – with changes in the economy, the social structure, the system and, above all, the challenges, the resistances to that system (Sivanandan 1983, p.2).

Briefly defined, racism refers to the assumption, consciously or unconsciously held, that people can be divided into a distinct number of discrete 'races' according to physical, biological criteria and that systematic social differences automatically and inevitably follow the same lines of physical differentiation. By this definition, the belief that black people are inherently more musical or more athletic than white people, for example, is a racist belief. It is also important to recognize that racist ideologies intersect in complex ways with other ideologies such as those of gender and class. Challenging racism in practice, therefore, simultaneously involves an engagement with patriarchy and capitalism and with the different forms that those structures assume in different historical and geographical circumstances.

The definition of racism advanced here can be extended to include the belief in inherent cultural difference, as I have argued elsewhere (Jackson 1987). The belief that Irish people are predisposed to physical aggression or to excessive drinking is no less racist than the previous examples about black musicality and athleticism. In each case it is the assumption of a direct and immutable link between biological or other inherent features and particular social characteristics that justifies the accusation of racism. Racist beliefs turn to the evidence of nature rather than history to explain perceived social differences. This chapter will therefore speak of *the idea of race*, rather than of race *per se*, to emphasize its socially constructed as opposed to its biologically given character.

This chapter explores the social construction of 'race' via an analysis of the language through which racism is routinely expressed. The emphasis on language derives from two specific sources. The first is Gareth Stedman Jones' *Languages of class* (1983) in which he argues that the word 'class' has acted as a congested point of intersection between many competing, overlapping, or simply different forms of discourse – political, economic, religious and cultural (ibid. p.2). This chapter suggests that there are many significant parallels with the idea of race and with the changing vocabulary of racism in recent British history. As with class, the idea of race is embedded in the English language and should be analyzed in its linguistic and social context. For there are different languages of race and class in different historical periods and geographical settings. The second source of inspiration for this chapter is Stuart Hall's observations on the nature of British racism (Hall 1980b) where he shows how contemporary ideologies draw on and modify a *reservoir of racist imagery* established in the course of four centuries of slavery and Empire. While contemporary racism is not simply an historical remnant of colonial stereotypes but an active creation that varies with present circumstances, there is nonetheless an important sense in which contemporary racist ideologies employ a *pre-formed vocabulary*, adopting and adapting an already available language – a repertoire of racist images and stereotypes that are drawn on selectively as occasion demands.

Most of the examples in this chapter will be drawn from the history of British (or, more specifically, English) racism.[1] That similar analyses can be undertaken for other capitalist societies with comparable, though distinct, historical trajectories is confirmed by recent Canadian research.[2] Kay Anderson, for example, has shown how the social construction of 'race' and the process of racial categorization can be translated directly into the politics of place. The idea of 'Chinatown' (Anderson 1987) is part of a white European tradition which has been projected on to the urban landscape in Vancouver and other Canadian cities and which can be read as an exercise in cultural hegemony (Anderson, 1988) involving the power of the state and the institutions through which that power is channelled. The social definition of 'race' is mediated, in this case, through the power of place and can be traced in the evolution of a distinctively political landscape.

Before proceeding to apply these ideas to the various forms of contemporary British racism, it is necessary to provide a brief history of the changing material context in which these ideas emerged.

The roots of British racism

It is often assumed, though incorrectly, that the history of British racism dates back no further than 1945 when large-scale immigration from the 'New Commonwealth' began.[3] Such a 'foreshortened historical vision' (Lawrence 1982a) fails to recognize that racist attitudes pre-date the arrival of a substantial black population in Britain at the end of World War II. A glance at the history of British attitudes towards Irish or Jewish immigrants in the 19th century is enough to dispel this myth.[4] It also implies that Britain's race-relations problems are attributable solely to New Commonwealth migration without considering the persistent hostility that British people have shown to 'foreigners' from every part of the world over several generations. The political implications of this misunderstanding of British history can be severe, justifying the imposition of strict immigration restrictions and legitimating the introduction of tightly defined nationality laws that many people regard as racist in their effects if not in their intent. An understanding of the history of British racism must therefore begin much earlier than 1945.

There has, in fact, been a black population in Britain at least since Shakespearean times. The reception of Shakespeare's *Othello* by Elizabethan audiences, for example, was no doubt affected by contemporary calls for the expulsion of 'divers blackamoors' of whom the Queen had declared in 1596 there were already here 'to manie' in a letter to the Lord Mayor of London (Fryer 1984, GLC 1986). Following a series of bad harvests, the price of bread had risen dramatically and the black population represented a convenient scapegoat for society's economic ills. This is just the first of many recorded instances when the black population has served a similar convenient function.

Throughout the 18th and 19th centuries, travellers' reports from diverse parts of the growing British Empire furnished descriptions of foreign lands and their 'alien' peoples which were read with considerable interest by the educated public at home in England. In 1788, readers of *Gentleman's Magazine* were treated to the following oft quoted account in which racist and sexist imagery vie for prominence:

> The Negro is possessed of passions not only strong but ungovernable; a mind dauntless, warlike, and unmerciful; a temper extremely irrascible; a disposition indolent, selfish and deceitful; fond of joyous sociality, riotous mirth and extravagant shew. He has certain portions of kindness for his favourites, and affections for his connections; but they are sparks which emit a glimmering light through the thick gloom that surrounds them, and which in every ebullition of anger or revenge, instantly disappear. Furious in love as in his hate; at best, a terrible husband, a harsh father and a precarious friend. A strong and unalterable affection for his countrymen and fellow passengers in particular seems to be the most amiable passion in the Negro breast . . .
>
> As to all the other fine feelings of the soul, the Negro, as far as I have been able to perceive, is nearly deprived of them (quoted in Walvin 1982, p.60).

Note that, for all its arrogance and condescension, there is a grudging respect for the Negro's 'warlike' disposition and a certain fearfulness for the 'riotous' quality of his (*sic*) mirth. The reference to 'joyous sociality' recurs some 20 years later in George Cruickshank's depiction of *Lowest life in London* (c.1805) (Fig. 6.1). Irrespective of their material deprivation, black people are shown contentedly dancing and singing in a public house. Similar images of black people as entertainers and domestic servants were a staple of William Hogarth's depiction of black people in 18th century London (Dabydeen 1985a).

The association between musicality and sexuality suggested by Cruikshank's illustration is a persistent feature of British racial discourse from 1750 to the present (Mair 1986). One particularly telling version occurs in the mid-19th century with the popularity of black-face minstrelsy. Troupes like the Ethiopian Serenaders (Fig. 6.2) helped perpetuate the idea of the comical, musical black. The special twist, though, in the case of minstrelsy is that the performers were not in fact black, but white, using burnt cork to blacken their faces and to portray

Figure 6.1 George Cruickshank's *Lowest Life in London*, 1821

Figure 6.2 *The Ethiopian Serenaders* c. 1840

Figure 6.3 *Songs of the Virginia Serenaders*, 1844

their idea of how black entertainers should behave in front of exclusively white audiences (Fig. 6.3).

During the 19th century, racial attitudes began to harden as all remaining traces of the 'noble savage' stereotype gave way to the image of 'comic buffoon' (Mair 1986). The content of the minstrel shows reflected these changes as sentimental songs were replaced by performances with coarser lyrics and harsher images of black people. Deteriorating attitudes reflected changing material circumstances both at home and abroad. The Morant Bay rebellion in Jamaica in 1865, brutally suppressed by Governor Eyre, led to increased fears about insurrection at home in Britain. Fears about street violence grew to hysterical levels and attitudes to race went through a similar period of crisis (Pearson, 1983, Bolt 1971).

Ideologies of class and race increasingly overlapped as the metaphor of 'darkest England' reflected fears of social unrest at home and abroad (see Ch. 4). Comics like *Boy's Own* were extremely popular reading for the urban middle classes, depicting the bravery of solitary white explorers overseas, confronting the 'yelling masses' of irrational, violent, angry natives (Fig. 6.4). Adventure stories like Rider Haggard's *King Solomon's Mines* (1885) and Edgar Rice Burroughs' *Tarzan of the Apes* (1917) perpetuated their own versions of white supremacy and a kind of casual racism that is still current in James Bond and Harrison Ford movies.[5]

As a discipline, geography was not immune to the influence of Empire and played a significant role in the development of racist attitudes in Britain. The professional origins of geography as a 19th-century 'European science' (Stoddart 1986) ensured that it was thoroughly immersed in the intellectual and political currents of the day including both social Darwinism and imperialism. As Harvey has written: 'Geographical practices were deeply affected by participation in the management of Empire, colonial administration, and exploration of commercial opportunities. The exploitation of nature under capitalism evidently often went hand in hand with the exploitation of peoples' (Harvey 1984, p.3). The links between Geography and Empire were sometimes very crude. For example, Bartholomew's *School Economic Atlas*, published in 1921, carried a map of 'The Races of Mankind'. The text accompanying the map was written by a Fellow (and Victoria Gold Medallist) of the Royal Geographical Society. Adopting the three-fold (black, white, yellow) racial classification proposed by de Gobineau, the atlas taught generations of school children the economic importance of race. Such passages as the following were commonplace:

> . . . in the case of the Negro, climactic influences – acting direct and through the typical food – lead to the early closing of the 'seams' between the bones of the skull; and thus the development of the brain is arrested; and the adult is essentially unintellectual. On the other hand, he is naturally 'acclimatized' against numerous diseases and other conditions of life and work which are very adverse to the white man. *He is, therefore, of great use as a manual labourer in a 'steamy' climate, e.g. on a cane-sugar plantation* (quoted in Cashmore & Troyna 1983, p.213; emphasis added).

Even after World War II articles about 'Social geography and its place

I emptied my revolver among
the yelling mass

Figure 6.4 Stereotypes of 'race' in early nineteenth-century popular culture

in colonial studies' could still be found (Gilbert & Steel 1945), while the racism and bias of geography textbooks continues to be a subject of serious concern (Hicks 1981).

Representing 'race'

Despite the continuities in British representations of race that this brief history suggests,[6] there was a dramatic shift in the mid-1940s coinciding with a relatively brief period of sustained migration from the Caribbean. At the peak of the migration in 1961 (before the Commonwealth Immigrants Act came into effect in 1962) some 50000 people arrived from the West Indies in one year. The migration was, in fact, quite sensitively attuned to labour demand in Britain (Peach 1968), partly because of direct recruitment from the Caribbean by London Transport and the National Health Service, and partly from the effects of information flows and chain migration.

Nonetheless, the most common representation of black people in the British press was of a *problem group*. The news magazine *Picture Post* ran a series of articles with titles like '30000 Colour Problems' and 'Would You Let Your Daughter Marry a Negro?'. These stories often had a liberalizing intention, focusing on the 'respectability' of most Caribbean migrants and on their relatively docile attitude towards manual labour: 'a useful job . . . and no one objects' (Fig. 6.5). They recognized that discrimination occurred in all but a few spheres of employment, like sport and entertainment, and argued that segregated communities represented the most permanent and symbolic obstacle to social progress. Yet the discourse of race they employed was cast in terms of personal prejudice and individual discrimination ('strange fogs of ignorance and prejudice') rather than in terms of racism and inequality ('Is There a British Colour Bar?', *Picture Post* 2 July 1949).

Stuart Hall, who has researched the images employed in *Picture Post* during the 1950s, argues that the selection of photographs to illustrate these pieces was far from innocent (Hall 1984). In the article about interracial marriage referred to above, for example, the lead picture shows a lone white woman with a baleful-looking black child. Although other photographs had been taken at the time, including shots of the black father playing contentedly with the child or out shopping with the family (Fig. 6.6), it was a picture of the lone mother and her child that was ultimately selected as the most appropriate, perpetuating the idea of a social problem whatever the explicit intention of the story.

As tensions between black and white people increased during the

Figure 6.5 'A useful job . . . and no one objects'

1950s, culminating in 'race riots' in Nottingham and Notting Hill in 1958 (Miles 1984), the association of black people with social problems became almost automatic. Salman Rushdie has traced this association and comments:

> The worst, most insidious stereotype . . . is the characterisation of black people as a Problem. You talk about the Race Problem, the Immigration Problem, all sorts of problems. If you are liberal, you say that black people have problems. If you are not, you say they are the problem. But the members of the new colony [black people in Britain] have only one real problem. That problem is white people. Racism of course is not our problem. It is yours (quoted in Bhavnani & Bhavnani 1985, p.146).

As the economy began to falter after the boom years of the 1960s, increasingly urgent calls were made by members of both major political parties to put a limit on further immigration, resulting in the Commonwealth Immigration Acts of 1961, 1968, and 1972. The liberal, permissive tone of public debate about immigration gave way to a harder language epitomized by Enoch Powell's rhetorical outbursts about the inevitability of 'racial' conflict. In speeches throughout the 1960s in his West Midlands constituency and elsewhere, Powell focused on two principal issues: the absolute numbers of 'coloured immigrants' and their propensity to increase more rapidly than the indigenous population, and their geographical concentration as an 'alien wedge' in British cities where they were held to be 'unassimilated and unassimilable' (Powell 1978).

Although Powell was ostracized in Westminster for his remarks and forced to resign his ministerial post, in retrospect it can be seen that his intervention radically altered the nature of British racial discourse, *making racism respectable*. The vocabulary and rhetorical form of Powell's infamous addresses now resonate throughout mainstream political debate without drawing the hostile attention that greeted Powell's notorious speeches.[7] For example, the same metaphors of 'floods and swamps' that Powell introduced into British racial discourse can clearly be seen in Margaret Thatcher's pre-election speech in 1979:

> If we went on as we are, then by the end of the century there would be 4 million people of the New Commonwealth or Pakistan here. Now, that is an awful lot and I think it means that people are really rather afraid that this country might be swamped by people with a different culture. And, you know, the British character has done so much for democracy, for law, and done so much throughout the world that, if there is a fear that it might be

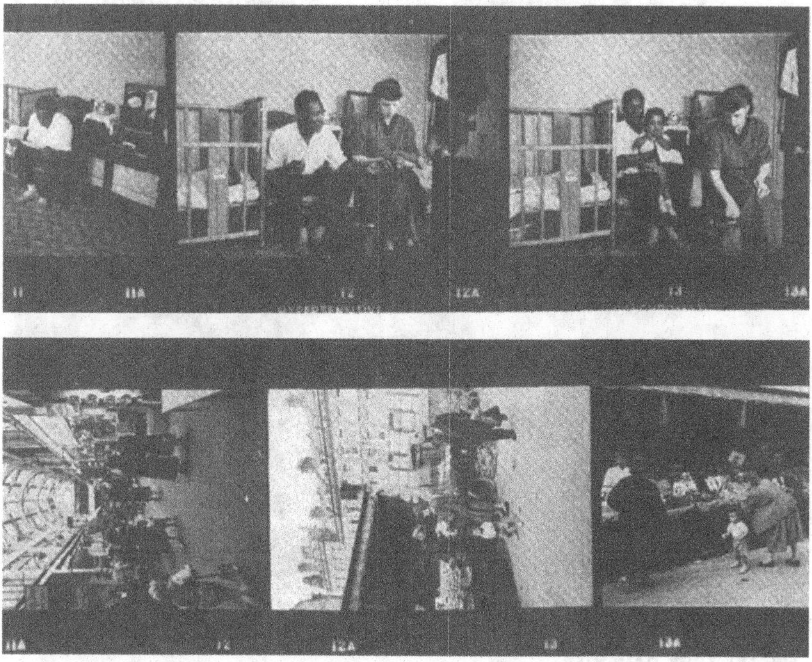

Figure 6.6 'Would you let your daughter marry a Negro?'

swamped, people are going to react and be rather hostile to those coming in (quoted in Sarup, 1986, pp.125–6).

The metaphor of 'floods and swamps' recurred again in press accounts of the visa controversy in October 1986 when the government introduced a system of compulsory visas for visitors to Britain from selected Commonwealth countries. Predictably, there was a rush to beat the imposition of this new legislation and airport immigration staff were, for a few days, unable to cope with the increase in people seeking admission to Britain. Press reports of the episode are interesting for a number of reasons. Blame for the administrative chaos that ensued was almost invariably laid at the door of the would-be 'immigrants' as, for example, in the *Daily Mail*'s headline: 'Immigrants Paralyse Heathrow' (15 October 1986). The same issue carried a picture with the caption: 'Swamped: immigration officer is beseiged by relatives seeking news of passengers'. The *Times* picked up the same metaphor with the headline: 'Heathrow under seige by Asians' (15 October 1986), while the *Sun*

THEY SAY IT LOUD ENOUGH FOR HER TO HEAR. "LOOK AT THAT, MAY. AIN'T IT A SHAME." It's the middle-aged women
worst, although the working man can be cruel, too. Near here there used to be whitewash on the walls. "Keep Brixton white." This girl has been
three years and has two children. Her husband is unemployed. He says that work is hard to find in Britain—especially for a Negro
YOU LET YOUR DAUGHTER
ARRY A NEGRO?

Figure 6.6 contd. 'Would you let your daughter marry a Negro?'

chimed in with '3000 Asians Flood Britain' (15 October 1986). But the *Daily Express* excelled them all with: 'Asian Flood Swamps Airport' (15 October 1986). It continued its story: 'Heathrow airport was under seige early today after a mass invasion of illegal immigrants trying to beat the midnight deadline for getting into Britain'. Its use of 'seige' and 'mass invasion' are characteristic of a whole genre of racist reporting (Cohen & Gardner 1982). But the reference to 'illegal immigrants' is entirely gratuitous as the legal status of those seeking entry had still to be determined.

Moreover, the word 'immigrant' itself is used here as a synonym for 'black' or 'coloured' despite the fact that the majority of British 'immigrants' are in fact white. Finally, it is worth noting that only the *Guardian* saw fit to interview a representative of any of the immigrant welfare organizations. All the other papers relied for their coverage of events on information supplied by politicians, immigration officials, airport staff, and security forces. Only the *Guardian*, again, implied in its headline, 'Heathrow Trap For Thousands in Visa Chaos' (15 October 1986), that the authorities bore some responsibility for the chaos that the new restrictions imposed.

Reporting the riots

The visa crisis was, except for those directly involved, a relatively minor episode in the history of postwar British race relations. Much more significant in their effect on British racial discourse were the urban 'riots' that broke out in several British cities during the 1980s, starting in Bristol in 1980 and spreading to London, Liverpool, and Manchester in 1981, and to Birmingham and elsewhere in 1985. Although only a minority of those arrested for riot-related offences were in fact black (Peach 1985), the riots are generally remembered as having been in some sense 'racial' (Gilroy 1987). The way the press reported the riots therefore, gives a further indication of the cultural politics of British racism.

Some newspapers were quite explicit in their attribution of a racial motive, even in Toxteth on Merseyside where relatively few black people were involved. The *Daily Express* had the most explicit reference of this kind in its headline: 'Black War On Police' (6 July 1981). References to violent conflict were common in other papers, where 'battles' raged across 'frontline Britain'. But there was also a more subtle stereotyping of these events as Jacquie Burgess reveals in her analysis of the way the media reported the 1981 riots (Burgess 1985).

Burgess suggests, in fact, that the press located the riots in a mythical 'inner city', an imaginary space that bore little resemblance to the very diverse circumstances of Brixton, Toxteth, and Moss Side. But this fictitious creation of a uniform and alien 'inner city' enabled the papers to report the violence as far removed from the everyday experience of the majority of middle-class readers and, hence, as none of their responsibility. Burgess argues that this mythical inner city was made up of four ideological domains: a physical environment of dilapidated houses, disused factories, and general dereliction; a romanticized notion of white working-class life with particular emphasis on the centrality of family life; a pathological image of black culture; and a stereotypical view of street culture.

Though few newspapers were prepared to identify a directly racial motive for the violence in 1981, they employed several key phrases, including references to the inner city itself, with distinctly 'racial' connotations. Indeed, one consequence of the disorders was to make 'riot' a virtual synonym for 'race'. Whereas in 1976, for example, the *Daily Telegraph* had led its story on the disturbances in Notting Hill with the headline 'Carnival Ends in Race Riot' (31 August 1976), by 1987 they described the recurrence of similar violence without any direct reference to race: 'Riot Police Battle With Mobs in Notting Hill' (1 September 1987). In this and other cases, particular place names are sufficient to connote the propensity for violent conflict. The second outbreak of violence in Brixton in 1981 gave rise to headlines such as: 'Brixton Explodes Again' (The *Sun* 11 July 1981). Also, in both Brixton and Notting Hill, several writers spoke of the 'tinder box' atmosphere, implying their vulnerability to any 'spark' or 'trigger', a set of metaphors that Lord Scarman later employed in his official inquiry into the Brixton disorders (Scarman 1981). This is an aspect of British racial discourse that deserves further analysis.

In general, however, as Michael Keith (1987) has shown, the 'riots' were simply appropriated by different interest groups and interpreted in extremely diverse ways with little concern for empirical evidence. Like football hooliganism (see Ch.3), the empirical void creates a space where writers from different ideological perspectives can project their preferred reading of events without fear of being contradicted by the evidence. Thus, while some journalists saw the riots as an indication of a more general urban distress ('Save Our Cities', *Daily Mirror* 7 July 1981), others appeared to share Margaret Thatcher's view that the riots were simply a 'spree of naked greed' and 'criminal violence' ('Riot Fury of Maggie: no excuse for this criminal violence', *Daily Express* 14 April 1981).

Confirmation of Hall's argument about the *pre-formed* quality of racial discourse is provided by striking similarities in the language used to describe the violence of 1976 in Notting Hill, and the violence in Merseyside in 1981, despite the very different nature of the two events. Responding to the prevalence of looting by children during the Toxteth disturbances, the *Daily Mail* asked: 'Don't Their Parents Care?' (8 July 1981), closely reflecting the ideas of a correspondent in the *Evening News* in 1976 who thought that 'West Indian parents for the main part have lost control over their children' (1 September 1976). The mythical structure of this kind of reporting and its evocation of imaginary places also demand closer analysis in any reconstituted cultural geography.

The myth of 'Asian' ethnicity

So far, most of this chapter has been concerned with Afro-Caribbean stereotypes. A parallel and contrasting set of stereotypes of 'Asian' minorities in Britain can also be identified. The first thing to be said, of course, is that 'Asian' is a highly contested term which has taken on political significance recently as an alternative to a broader Black British identity. In Britain, 'Asian' is normally taken to refer to people from the Indian subcontinent (from India, Pakistan, Bangladesh, and Sri Lanka) and their British-born descendants.[8] But it may also include those from China, Japan, and elsewhere in South East Asia. The internal diversity of the 'Asian' population, even excluding those from South East Asia, is sufficient to make the term of rather dubious value in the British context. There is no uniform Asian life-style or culture, for example. Differences of nationality, religion, and language suggest that the only thing Asian people have in common is being regarded as members of a single ethnic group by the majority society. Until recently, Asian people in Britain have rarely described themselves by this label and only do so now to distinguish themselves from other black groups such as the Afro-Caribbeans with whom they are in competition for resources. The debate around these issues concerns the degree to which internal divisions are relevant in a racist society such as Britain which tends to impose a similar experience on all black people. In this sense, 'black' (or Black) refers not so much to skin colour as to a state of political consciousness, acknowledging the common experience of racism and a willingness to unite in challenging it. While 'Black' can be defended politically, as primarily a matter of consciousness rather than colour, it is freely admitted that not all black people share this usage.

Similar problems surround the use of the term 'Afro-Caribbean', preferred here to 'West Indian' with its colonial implications. The Office of Population Censuses and Surveys, which has been experimenting with various different ways of asking an 'ethnic question' in 1991 (OPCS 1987), rejects 'Afro-Caribbean', saying that the majority of older, overseas-born, 'West Indians' prefer that term. The Census continues to have problems classifying persons of 'West Indian' descent who were born in Britain (among whom the term 'Afro-Caribbean' is becoming more popular). The more fundamental problem remains, however, as significant numbers of people object to *any* version of the ethnic question.

Such terminological sophistication is of little relevance in defining popular attitudes to Asian people where a cruder set of stereotypes exists to prevent exactly this kind of ambiguity. Here, the word 'Asian' tends to conjure up a constellation of social attributes such as 'crafty', 'sneaky', and 'mean', including a grudging respect for their reputation as hard workers. Their alleged success in business, however, is held to have been won at an intolerable cost in terms of sociability and *joie de vivre*.[9] Aside from these implicitly male stereotypes, a parallel set of images exists for Asian women, stressing their alleged passivity, dexterity, and docility, coupled with the erotic appeal of their 'exotic' sensuality (Parmar 1984). Airline advertisements to the Far East often trade blatantly on these associations.

There are some interesting contradictions within these stereotypes which reflect their historical roots and geographical origins. The patronizing colonial attitude that all Asians were alike runs counter to the Victorian passion for classification and subdivision. Thus, in 1878, the Oxford professor of Sanskrit could write confidently of the 'racial diversity' of the Indian population, while at the same time betraying a condescending attitude to them all. He identified the following range of characteristics: the spirited Hindustani, the martial Sikh, the ambitious Marathi, the proud Rajout, the hardy Gurkha, the calculating Bengali, the busy Telugu, the active Tamil and the patient Pariah (quoted in Bolt 1971, p.186). Even where these stereotypes appear to bestow approval (as with the 'West Indian's' alleged propensities for sport and entertainment), they serve a restrictive and exclusionary function, confining black people to certain occupations and reflecting an inequality in the power of definition which is not afforded reciprocally to those who are thus defined.

Similar contradictions are present in contemporary racial stereotypes. Derogatory views of Asian people, for example, are countered by a reluctant respect for the 'historical depth' of their cultures. This is often contrasted favourably with the supposed absence of 'cultural depth' in Caribbean societies where generations of migration, slavery, and colonial rule are held to have obliterated all traces of an authentic indigenous culture. References to the strength and depth of Asian culture are no less racist than the denial of these qualities in the Afro-Caribbean context. Recent government reports are full of this kind of stereotyping, however, with dire consequences for members of both groups. Lord Scarman, for example, described 'the young people of Brixton' as 'a people of the street' (Scarman 1981, p.11), criticizing 'West Indian children' for their 'educational under-achievement' and their parents for a lack of 'active interest in the education of their children' (ibid. p.106). The effects of similar stereotypes have also been fiercely debated in the context of 'multi-cultural' education where the 'under-achievement' of Afro-Caribbean children is often contrasted with the exemplary performance of Asian pupils, without giving sufficient consideration to the self-fulfilling nature of teacher expectations (cf. Swann 1985, Sarup, 1986). The contradictory nature of these stereotypes can also be traced historically by reference to Edward Said's penetrating analysis of *Orientalism* (1978).

Imagining the Orient

Said's book deals with the complex and contradictory attitudes of Western society towards the 'Orient', exploring how Europeans are able to admire and respect Arab and Islamic cultures as the source of European languages and civilization while at the same time holding demeaning and derogatory stereotypes about 'Orientals'. Said suggests that the Orient provides Europeans with their deepest and most recurring sense of the 'Other'. Attitudes towards it are therefore often expressed through a series of oppositions which take an idealized version of European civilization as their implicit point of reference. Said cites Lord Cromer's high-handed attitude towards Egypt as a typical instance of this mentality:

> The European is a closer reasoner; his statements of fact are devoid of any ambiguity; he is a natural logician, albeit he may not have studied logic; he is by nature sceptical and requires proof before he can accept the truth of any proposition; his trained intelligence works like a piece of mechanism. The mind of the Oriental, on the other hand, like his picturesque streets, is eminently wanting in symmetry. His reasoning is of the most slipshod description. Although the ancient Arabs acquired in a somewhat higher degree the science of dialectics, their descendants are singularly deficient in the logical faculty. They are often incapable of drawing the most obvious conclusions from any simple premises of which they may admit the truth. Endeavour to elicit a plain statement of facts from any ordinary Egyptian. His explanation will generally be lengthy, and wanting in lucidity. He will probably contradict himself half-a-dozen times before he has finished his story. He will often break down under the mildest process of cross-examination (quoted in Said 1978, p.38).

Said's most radical insight, and the point he develops at length, is that the 'Orient' was almost a European invention: a place of romance, exotic beings, haunting memories and landscapes, and remarkable experiences. The European sense of self is, in part, premised on the existence of another, different world, largely a product of its own imagination. East assumes West; the Orient assumes the Occident. The two geographical entities support and reflect each other (ibid. p.5).

Said is at pains to point out that the relationship between East and West is not a purely imaginative relation but one that is based on very real material foundations, including the history of French, British, and American imperialism. Indeed, the Orient provided Europe's oldest and richest colonies and has been its persistent cultural contestant. For this reason, the bulk of Said's analysis is taken up with a consideration of 'Orientalism' which he defines as the discourse through which the West has sought to legitimize its dominance and authority over the Orient. Orientalism, then, is not a 'myth' in the sense of a demonstrable falsehood; it is more powerful that that. It is a body of theory and practice in which there has been considerable material investment over many generations.

Said shows how Orientalism has been subjected to every Western ideology from imperialism, positivism, utopianism, and historicism, to Darwinism, racism, Freudianism, Marxism, and Spenglerism, but that ultimately it has remained a *political vision* the structure of which prompted the difference between the familiar (Europe, the West, 'us') and the strange (the Orient, the East, 'them') (ibid. p.43). Thus, Orientalism represents an 'imaginative geography', a *textual universe* of academic and political ideas. Drawing on Bachelard's *The poetics of space* (1964), Said describes how this imaginative geography has helped to define the West by dramatizing the distance between what is close and what is far away, what is familiar and what is exotic. Through its imaginative geography, Orientalism imposes and draws upon a limited vocabulary and imagery (cf. Hall 1980b). More precisely, the imaginative geography of Orientalism:

> . . . legitimates a vocabulary, a universe of representative discourse peculiar to the discussion and understanding of Islam and of the Orient. What this discourse considers to be fact – that Mohammed is an imposter, for example – is a component of the discourse, a statement the discourse compels one to make whenever the name Mohammed occurs. Underlying all the different units of Orientalist discourse . . . is a set of representative figures, or tropes. These figures are to the actual Orient . . . as stylized costumes are to characters in a play (Said 1978, p.71).

It therefore follows that:

> . . . we need not look for correspondence between the language used to depict the Orient and the Orient itself, not so much because the language is inaccurate but because it is not even trying to be accurate. What it is trying to do . . . is at one and the same time to characterize the Orient as alien and to incorporate it schematically on a theatrical stage whose audience, manager, and actors are *for* Europe, and only for Europe (ibid. pp. 71-2).

Orientalism is a self-referential, closed system, effectively sealed off from the empirical world. Any reference to direct observation and experience not cast in the terms that Orientalist discourse compels would endanger the whole structure. It relies on its own peculiar logic:

> Orientalism staked its existence, not upon its openness, its receptivity to the Orient, but rather on its internal, repetitious consistency about its constitutive will-to-power over the Orient. In such a way Orientalism was able to survive revolutions, world wars, and the literal dismemberment of empires (ibid. p.222).

Finally, Said shows how the discipline of geography itself played a vital strategic role in the development of Orientalism, supervising the division of geographical space between the rival French and British colonial powers. For, despite their differences, both France and Britain saw the Orient as a single geographical, cultural, political, demographical, sociological, and historical entity over the destiny of which they believed themselves to have a traditional entitlement. Geographers are now themselves turning to Said's work in coming to terms with the power of ideology in defining social relations through space. Anderson's (1988) study of the development of Vancouver's Chinatown as an exercise in cultural hegemony is just one example of the fruitful application of Said's ideas in contemporary social and cultural geography.

Conclusion

Having spent so much of this chapter dealing with the world of representations – with racism as a cultural discourse – it is worth returning to the point made at the beginning about the intimate connection between ideology and practice. Racist ideologies have severe practical consequences particularly where they become institutionalized through the power of the state. For racism in Britain and similar societies is a *dominant ideology*, not just a matter of individual prejudice and personal discrimination. On the contrary, racism refers to a set of ideas and beliefs that have the weight of authority behind them; they are enshrined in statutes and institutionalized in policy and practice. In a capitalist society such as Britain, this implies that there is a close connection between race and class, mediated politically by the role of the state. Exactly how this relationship is to be theorized remains the subject of intense debate.

The problem concerns how best to deal with the multiple ways in which questions of class intersect with those of 'race' in the competition for scarce resources such as housing, education, and employment. All too often, 'race' is simply *reduced to class* as, by extension, it is argued that 'race' interferes with the development of class consciousness. Ida Susser's account of urban politics in the Williamsburg-Greenpoint district of Brooklyn in New York comes close to such a reduction, providing a sensitive account of gender relations but giving far less autonomy to 'race' which, she says, has 'interfered with the development of a sense of common purpose between groups' (Susser 1982, p.viii). Similarly, while admitting that under some circumstances 'race' may actually stimulate collective class activity, Ira Katznelson still speaks of the 'racial and ethnic fragmentation' of the working class (Katznelson 1981, pp.10-11). Even Richard Harris, who advances a thoughtful argument about the contradictory effects of residential segregation on class formation in capitalist cities does not extend his discussion to the issue of 'race' (Harris 1984).

Stuart Hall proposes, to my mind, the most satisfactory resolution to this theoretical problem, arguing that, for many blacks in Britain, 'class' is experienced *through 'race'* (Hall *et al.* 1978). While black people are exploited through class as well as through 'race', it is through the latter that they experience their prime sense of oppression. It is racism that sets the limits on their social actions, simultaneously comprising the structural determinant of their subordination and the medium through which they can most readily challenge that subordination. It is one of the paradoxes of 'race' that it defines the nature of black people's oppression while at the same time presenting a medium for the expression of resistance and opposition to that oppression (cf. Jackson & Smith 1981). Such a formulation is an advance on the idea that race and class form a 'double oppression' as this notion all too readily leads to the idea that the effects of race and class can be disaggregated empirically and treated in a simple additive way. Attempts to account for the relative effects of race and class on levels of residential segregation (e.g. Taeuber & Taeuber 1965) clearly fall into this trap and emphasize the need for an alternative theorization. A reconstituted cultural geography would be concerned to trace the circumstances in which particular combinations of race, class, and gender occur, and why each has a variable salience in different places, at different times.

This chapter has also provided tacit support for those, like Lawrence (1982b), who express a reluctance to speak of ethnicity as an independent dimension of social life with causal powers of its own. Lawrence rejects this notion of ethnicity as referring too quickly to the realm of culture, assuming the existence of a kind of pluralism that rarely occurs in practice. Instead, he would wish to see a broader recognition of the role of racism in defining the context of black-white relations, treating ethnicity as a series of more or less self-conscious strategies employed by subordinate groups to 'handle' or contest their structural subordination. This definition of ethnicity as a kind of political or cultural strategy is clearly much more consistent with the notion of cultural politics employed elsewhere in this book (cf. Gilroy 1987). As a term, 'ethnicity' may best be avoided insofar as it implies minority status without recognizing the centrality of power to the social relations implied by such a status. The case for retiring the concept of ethnicity is even stronger when it is simply used as a polite synonym for 'race'.

What, then, are the consequences of such a theoretical re-thinking for future research in cultural geography? One way of advancing the argument would be to clarify the role of 'race' within broader processes of economic and social restructuring, probing the effects of successive spatial divisions of labour on different sections of the population (Massey 1984). With few exceptions (e.g. Cross 1985), little progress has been made in this respect compared, for example, to theories on the rôle of class and gender. Such an analysis of the significance of 'race' in the spatial restructuring of society would help to clarify the process whereby racism is spatially as well as socially constituted (Jackson 1987).

The kind of empirical research that this theoretical approach might inform would start by examining the timing of different streams of migration and the social and economic consequences that follow from the location of different groups within successive spatial divisions of labour. It would proceed to explore the effects of this process on the selective 'racialization' of the labour force (as suggested by Miles 1982), allowing at least some speculation about why resistance has taken different forms at different times, focusing first on immigration issues at the point of entry, then on housing and employment, then on education and, most recently, on the streets. From such a basis it might be possible to speak of the *sites of struggle* in more than a metaphorical sense and to begin to describe the emergence of a *geography of resistance*.

Notes

1 Whether to speak of British or English racism is a complex political question. While the colonial experience was a British one, contemporary racism certainly takes a different form in Scotland than in England (Dummett 1973, Miles & Dunlop 1987). While recognizing the existence of different types of racism within Britain, they will be regarded here as varieties of a common British racism.

2 Certain forms of racism can, of course, be recognized in pre-capitalist societies. But, as Miles (1982) has shown, racism takes a different trajectory once it is associated with labour migration under capitalism.

3 The New Commonwealth includes the former British West Indies and various countries in the Indian subcontinent including India, Bangladesh, and Sri Lanka. Pakistan, though no longer part of the Commonwealth, is often included with the New Commonwealth countries for statistical purposes in order to generate a working definition of the black population in Britain. The term 'New Commonwealth' was coined to distinguish these countries from 'Old Commonwealth' countries like Canada and Australia the citizens of which are predominantly white.

4 Anti-Irish and anti-Semitic feeling can be regarded as forms of racism where they are rooted in assumptions about the biological basis of human behaviour. Thus it is 'racist' to assume that all Irish people have red hair or speak in a brogue. Similarly, anti-Semitic humour is racist where it refers to 'typical Jewish features' or to characteristics that are assumed to be innate.

5 Several recent studies have begun to examine the imagery of the popular fiction of this period. See, for example, Street (1975) and Dabydeen (1985b).

6 The next two sections include material that was published in abbreviated form in an earlier paper (Jackson 1988b).

7 A recent leader in *The Guardian* on 'Racism: the words and the reality' suggests that this may be so because 'race politics in Britain are increasingly conducted in a superficially non-racist code', while debates about the inner city, council housing, education, and law and order cannot be understood without appreciating their 'unspoken sub-text on race' (15 April 1987).

8 In the United States, 'Asian' refers principally to people from China, Japan, Korea, and the Philippines. Likewise in the US, 'Hispanic' is a collective noun that lumps together an extremely divers population from Mexico, Cuba, Puerto Rico, Central America, and elsewhere.

9 The contradictions that inhere within the myth of Asian business success have been explored by Cater & Jones (1987). They show that, contrary to the stereotype of the over-achieving Asian entrepreneur, most Asian businesses are concentrated in the declining corner-shop sector while most Asian-owned properties are in the most marginalized and deteriorating areas of the inner city.

Chapter seven
The politics of language

Cultural geography and linguistic theory

Among the social sciences, linguistics has often appeared to have made most progress in attaining the status of a true science. Since Saussure distinguished between *langue* (language) and *parole* (speech), identifying the phoneme as the basic building block of every natural language (Saussure 1916), linguistics has provided the model for a whole range of other structuralisms, from anthropology to psycho-analysis. In *Structural anthropology* (1963), for example, Lévi-Strauss makes some extravagant claims for the scientific status of linguistic analysis. It is, he says, probably the only social science which can truly claim to be a science having achieved the formulation of an empirical method and an understanding of the nature of its data (ibid. p.31).

Lévi-Strauss once compared the discovery of the phoneme to the discovery of the atom. If the 'elementary structures' of language could be scientifically identified, then the basic properties of other social systems, such as kinship and marriage, might also be uncovered. For all such systems, Lévi-Strauss argued, were but elaborations on a limited number of structural principles, such as the ancient incest taboo which marks the human divide between nature and culture (Lévi-Strauss 1969). According to Lévi-Strauss, particular kinship systems are part of a broader system of similarities and differences, each system comprising a particular permutation of a few basic themes. In providing similar analyses of totemism, mythical structures, and other symbolic practices, Lévi-Strauss gave considerable impetus to the extension of structuralist analysis beyond its origins in linguistic theory. Today, though, much of the optimism that surrounded these early breakthroughs in linguistic theory has been qualified as the social context of individual acts of speech has come once again to the top of the agenda. Among modern linguists, Roy Harris has championed this position, criticizing the idea that a linguistic community is no more than a congregation of talking heads:

> For language-making involves much more than merely the construction of systems of signs. It is also the essential process by which men (*sic*) construct a cultural identity for themselves, and for the communities to which they see themselves as belonging (Harris 1980, Preface).

As elsewhere in the social sciences, such forms of 'grand theory' are on the retreat, any form of 'social physics' is regarded with suspicion, and the need to contextualize the study of language has gained widespread recognition (cf. Williams, 1988). This chapter therefore bypasses some important phases in the development of linguistic theory. It does not deal with the psycho-analytic theory of language, associated with the work of Jacques Lacan or with the cultural semiotics of Roland Barthes, neither of which provides an appropriate basis for a fully contextualized, historically grounded social geography of language and culture.[1] Instead, it concentrates on other developments in sociolinguistics besides those in semiotics, narrowly defined as the study of signs and symbols. It draws its inspiration from Foucault's approach to the analysis of discourse and from Marxist theories of ideology rather than from Barthes' semiology or Lacan's pyscho-analysis.[2]

These developments in linguistic theory have been virtually ignored by cultural geographers who have tended to restrict themselves to the analysis of patterns of spatial differentiation, mapping linguistic variation over space.[3] These preoccupations saved geographers from the pitfalls of an ungrounded structuralism, but they also isolated them from other advances in linguistic theory that bore directly on their empirical concerns. For much recent research in sociolinguistics would suggest that linguistic surfaces are in fact *continuous*, not subject to the kind of breaks and discontinuities required for simple cartographic representation. For example, the English spoken in Norfolk differs in a host of subtle ways from that spoken in Suffolk, suggesting a rich and highly variegated geography of language, built up from almost endless variations among speakers according to differences of age, gender, class, and context (cf. Trudgill 1975, 1983). Far from invalidating geographical work on language, however, this realization opens up a whole new range of possibilities for geographical research. This chapter reviews some of these possibilities.

The problematic character of language

As the discussion of Raymond Williams' *Keywords* (1976) in Chapter 2 suggests, words and concepts cannot be divorced from the social context in which they are used. Williams' contextual history of a selection of 'keywords' challenges the spurious authority of dictionary definitions which insist on a single, 'correct' meaning that artificially freezes the dynamic of linguistic change at an arbitrary point in time. Williams demonstrates that words like democracy, industry, class, art, and culture took on their current meaning during the period of intense industrial change in the 19th century. Some words virtually reversed their former meaning, such as 'individual' (which previously meant 'indivisible') or 'democracy' (which had a strongly negative connotation until the 19th century). Words are not just a passive reflection of their historical context, however. As Williams' work shows, many important social and historical processes take place *within language*. Gareth Stedman Jones (1983) provides several further examples, showing how concepts like class are embedded in their social and linguistic context. Stedman Jones suggests that experience cannot be abstracted from the language in which it is expressed; language structures and articulates experience, disrupting any simple notion of the unmediated determination of consciousness by existence.

This chapter explores some of these problematic aspects of language which simultaneously structures and reflects human experience and social action. These ideas have been popularized recently in the social theory of Anthony Giddens who uses the example of language as an illustration of the process of structuration. According to Giddens (1979, 1985), the structural properties of social systems are both the medium through which those systems are generated and transformed, and the outcome of a whole series of practices that constitute those systems. Thus, individual acts of speech draw on a pre-existing structure of grammar and actively reproduce it through the recurrent practice of grammatically correct speech. Systems of grammar have no independent existence; they represent the outcome of many individual acts of speech, sedimented through time into rules and conventions. Language is a structure of signification that is reproduced in social practice. Like other practices, however, it does not exist outside social relations of power. Grammatical rules and other linguistic conventions provide a system of sanctions through which certain practices are legitimized and social norms enforced (see Fig. 7.1). There is, in other words, a *politics of language*, just as there is with the other cultural systems and practices described in this book.

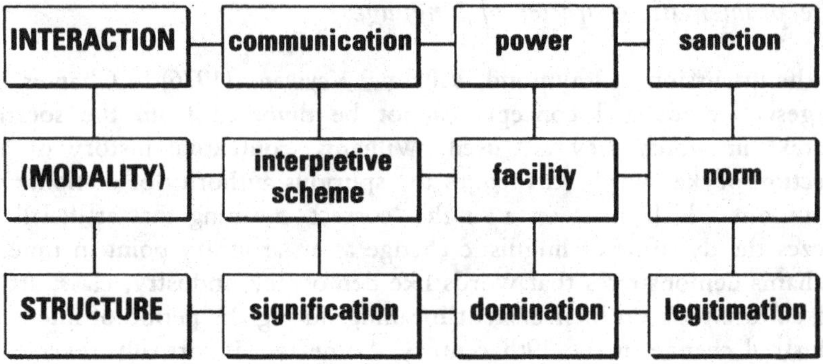

Figure 7.1 The theory of structuration

Linguistic variation: dialects, pidgins, and creoles

It is a truism that language varies across space and over time, but the precise contours of linguistic change are far from straightforward. An alternative to the kind of static, large-scale regionalization that has dominated cultural geography in the past is to begin by exploring the way that speech communities are built up, how they define and contest their boundaries in both a social and a spatial sense.

Trudgill (1983) cites the example of a casual encounter in a railway compartment where two English men strike up a conversation about the weather. He suggests that, unconsciously, neither of them may be particularly interested in the subject but that a casual conversation of this kind is a conveniently indirect way of seeking clues about status and background, comparable to the clues given off by appearance (dress, manner etc). Trudgill's example indicates the range of significant elements that may be present in even the most casual encounter: whether either party chooses to speak at all, who initiates the conversation, whether it is reciprocated, and so on, irrespective of the content of the dialogue and how it is delivered. But the extent to which inferences about *structure* can be drawn from the analysis of *interaction* is a fundamental problem for social science and one where linguistic theory provides only the roughest guide.

The men on the train in Trudgill's example employ a variety of social and spatial clues, conventionally defined in terms of class (accent) and locality (dialect). These terms are by no means uncontested, however. Whereas 'accent' is normally taken to refer to differences of pronunciation, 'dialect', in purely linguistic terms, includes differences of vocabulary and grammar as well as pronunciation. The study of dialect provides one of the best examples of the politics of language. For to refer to a dialect is to make a political rather than a strictly linguistic judgement. It involves an assessment of the relative merits of different types of language. A Yorkshire or Glaswegian accent is commonly held to be a variant of Received Pronunciation (the 'correct' way of speaking 'standard English'). Yet 'standard' or 'BBC English' is far from the common language of the majority of English speakers and, even as an ideal type, it is not universally accepted. In fact, it has been estimated that only about three per cent of the UK population actually speak 'the Queen's English', which Tom Nairn characterizes as a 'slurred, allusive, nasal cawing' (Nairn 1988, p.68). This 'dialect' is spatially as well as socially distinctive:

> Great Britain's accepted tongue is the ultra-distilled by-product of drawing-room, shoot and London club, a faded aristocratic *patois* remarkable for its anorexic vowels and vaporized consonants. It is social geography that links this vernacular to the London-Oxford-Cambridge triangle; while the social power of the same locality . . . has turned it into the inevitable emblem of authority, acceptance, literacy and nationality (ibid. pp.68-9).

Usually, then, dialect implies a debased or inferior kind of speech, or a 'minority' language (such as Catalan), as opposed to a 'majority' language (such as the Spanish of Castile). But what grounds are there for regarding one language or dialect as inferior to another, rather than treating them as separate languages of equal stature (such as Welsh in relation to English)? To return to a point made earlier in this chapter, linguistic surfaces are continuous: there are no clear breaks between different dialects and no linguistic rules for judging the status of different languages. If distinctions are made, then, they are made on political rather than on linguistic grounds: 'value judgements about language are, from a linguistic point of view, completely arbitrary' (Trudgill 1983, p.21). Different dialects attract different degrees of prestige, although not all groups within a particular linguistic community would rank each dialect in an identical fashion. Someone who speaks with a Yorkshire accent, for example, might not agree that standard English was a 'better' way of speaking and speakers of American English might not agree that British English is intrinsically superior.

The case of creole languages provides another good example of the politics of language and how it operates over space and time. Within Europe, for instance, several distinct languages are commonly recognized as having a high degree of autonomy (English, French, and German, for example). This distinction is maintained even though there is a good deal of variation (heteronomy) within each language and a fair degree of mutual intelligibility. Among English-speakers, however, various dialects may be distinguished, such as Afro-Caribbean creole or black English vernacular, which are not commonly recognized as autonomous languages. There is nothing *lingusitically inferior* about these forms of English. Like other fully-fledged languages, they are structured, complex, and rule-governed systems for conveying meaning among a community of speakers. Any judgements about the inferiority of creole reflects *social attitudes* towards the speakers of this dialect and perceptions of their social inferiority. Important political implications follow from judgements of this kind, concerning English-language teaching in 'multi-cultural' schools, for example, where it is commonly argued that Asian children have to be taught English as a second language while Afro-Caribbean children are assumed to speak a debased form of standard English.

But what exactly is meant by describing Afro-Caribbean speech as 'creole'? The word refers to the way that language changes as a result of human migration. When, as a result of migration, speakers of two distinct languages come into contact, communication often takes place by means of a *lingua franca*. In many cases, the *lingua franca* is not indigenous to the area where it is spoken (as when English became the *lingua franca* in many parts of the British Empire). In other cases, particularly in Africa, the *lingua franca* was indigenous (as with Swahili). The development of a *lingua franca* often involves the simplification of a more complex language which may become relatively stabilized to form a pidgin language (a *lingua franca* that has no 'native' speakers but which has to be learnt). Over time, as children learn pidgin from their parents, a process of *creolization* occurs. Creole languages, then, are pidgins that have acquired native speakers (Trudgill 1983, p.182). But, as Trudgill argues, they are perfectly normal natural languages. There is absolutely nothing 'wrong', linguistically, with Afro-Caribbean creole. Any judgement about its inadequacies or about its status in relation to 'standard English' is therefore a political, not a linguistic, judgement.

Language also changes as a result of changing social contexts. Words change their meaning as they move from one sociolinguistic domain to another (from slang to everyday speech, for example, or from one speech community to another). The extent to which this kind of linguistic dynamism is spatially defined is relatively unexplored (though see Trudgill 1982). Clearly there is a regional dimension to linguistic change, as dialect words enter the linguistic mainstream or as 'minority' languages persist in geographically marginal areas. But there is no simple correlation between social and spatial change. As Aitchison and Carter (1987) have demonstrated, for example, there has been a distinctive resurgence in the number of Welsh speakers in Cardiff. But, contrary to Hechter's (1975) model of internal colonialism, it has taken place among upwardly-mobile professional people rather than among more disadvantaged classes, and in cities rather than in remoter rural areas. In order to develop these ideas about the social dimensions of linguistic change, it is necessary to elaborate on the concept of linguistic communities introduced in this section. For, as Duncan and Duncan (1988) have argued, social groups can be construed as *textual communities* as well as speech communities. Textual communities form around shared readings which reinforce a group's identity and mark it off from neighbouring communities. If the textual analogy is applied to landscapes, as in Duncan and Duncan's work, then 'reading' the landscape becomes a thoroughly political process.

Linguistic communities

In an important paper on language and subjectivity, Harrison and Livingstone (1982) highlight the centrality of language in structuring people's subjective experience and in rendering what is experienced personally interpretable by a wider community. Language, they argue, is significant as the principal medium through which intersubjective meaning is communicated, playing a crucial role in structuring people's social and cultural identities (ibid. p.7). Communication takes place within *linguistic communities*, characterized by the possession of shared belief systems, myths, and ideologies, as well as a common language. This need not imply that whole societies share a single language, for every society will contain many linguistic communities, defined by age, gender, class and so on. But the analysis of language provides a key point of entry into the analysis of social distinctions. The structuring of language into systems of dominance and subordination, as described by Giddens' theory of structuration for example, provides a way of understanding how the negotiation of meaning between groups becomes sedimented into more permanent structures and relations of inequality. Foucault (1980) makes a similar point by describing the way that discourses develop around particular ideas, reflecting and reproducing existing power relations.[4]

The broader question of language and the communication of meaning has been taken up by George Steiner (1975) who suggests that any act of understanding requires an act of *translation*. As a literary critic, Steiner begins by showing that any thorough reading of a text out of the past of one's own language and literature is a manifold act of interpretation. For example, he suggests that Jane Austen's deceptively urbane prose represents a radically linguistic world in which reality is 'encoded' in a distinctive idiom: what lies outside the code lies outside Jane Austen's criteria of admissible imaginings (ibid. p.9). Decoding its literary and social meaning requires multiple layers of interpretation and translation.

There are 'worlds within worlds' in *Mansfield Park* (1814), for example, reflected in the linguistic forms employed. The novel moves between various places (London, Portsmouth, and Mansfield Park), each representing a different social world. The arrival of Mary and Henry Crawford from London threatens the peace and security of Mansfield Park as Mary tries to dissuade Edmund from being ordained and Henry, her brother, flirts with the novel's heroine, Fanny Price. The Crawfords make the most of Sir Thomas Bartram's temporary absence from Mansfield Park, 'relieved by it from all restraint'. At a crucial point in the novel, Mansfield Park becomes the setting for amateur theatricals, giving Henry and Mary Crawford a further pretext for flirtation and irresponsibility. Significantly, Fanny does not wish to take part: 'I could not act any thing if you were to give me the world'. Sir Thomas' timely return prevents further consternation and re-establishes the peace and order of the house. The novel conveys its moral purpose as much by language as by plot, by the juxtaposition of words and phrases that convey the opposition between conflicting social worlds: manners and morals, personality and principle, wit and wisdom. From this point of view, the 'conservative' tone of Jane Austen's prose is a perfect vehicle for her social criticism, conveying both the restrictions and the possibilities that exist within a socially and spatially encapsulated world.

A further example of the use of linguistic form to symbolize the opposition between moral worlds is provided by the poetry and novels of the so-called 'Beat Generation' (Jack Kerouac, Allan Ginsberg, and others). For these American writers in the 1950s, constant movement was an end in itself. Offering a geographical reading of their work, Tim Cresswell suggests that 'mobility' represents a challenge to the moral authority of 'place', a geography of resistance to the idea of 'rootedness' conveyed in linguistic rather than spatial terms (Cresswell 1988). He traces this tension in the work of Kerouac and others who argued that:

> In America camping is considered a healthy sport for Boy Scouts but a crime for mature men who make it their vocation. Poverty is considered a virtue among the monks of civilized nations – in America you spend a night in the caboose if you're caught without your vagrancy change (Kerouac 1960, p.174).

Kerouac developed the theme of mobility in his autobiographical novel, *On the road* (1958), where his staccato narrative style and the episodic nature of the action convey an urgent sense of dissatisfaction with the humdrum world of conventional morality:

> Suddenly I found myself on Times Square. I had traveled eight thousand miles around the American continent and I was back on Times Square; and right in the middle of the rush hour, too, seeing with my innocent road-eyes the absolute madness and fantastic horror of New York with its millions and millions hustling forever for a buck among themselves, the mad dream – grabbing, taking, giving, sighing, dying, just so they could be buried in those awful cemetery cities beyond Long Island City (Kerouac 1958, p.106).

As Cresswell shows, the road is associated with innocence, holiness and purity; the city with madness, nonsense, and confusion. Mobility symbolizes freedom from a materialistic world, dominated by the pursuit of money and bourgeois respectability.

Radical differences between linguistic communities also affect the critical appreciation of contemporary literature. Cora Kaplan, for example, suggests that a reading of Alice Walker's *The Color Purple* (1983) demands a substantial act of the imagination on the part of white, middle-class readers if they are to avoid an 'ideological bleaching' of the text (Kaplan 1986, p.178). A literary analysis of Walker's text, she argues, must be developed from an understanding of the specific histories of race, class, and gender in the American South as well as an appreciation of the precise cultural moment in which the book was written, published, and read.

The social geography of linguistic communities in the Middle Ages has been explored in a thoroughly innovative way by the Soviet literary critic, Mikhail Bakhtin. In his highly original analysis of *Rabelais and his world* (1984), Bakhtin shows how the language of the market-place functions as a critique of bourgeois conventions, inverting and subverting the official moral order of the court. Every act of world history, according to Bakhtin, has been accompanied by a laughing chorus (ibid. p.474). In carnivals, fairs, and other popular entertainments, the official order of bourgeois respectability was satirized symbolically and ritualistically, above all through language:

> . . . the unofficial culture of the Middle Ages and even of the Renaissance had *its own territory* and its own particular time of fairs and feasts. This territory . . . was a peculiar second world within the official medieval order and was ruled by a special type of relationship, a free, familiar, marketplace relationship. Officially, the palaces, churches, institutions, and private homes were

> dominated by hierarchy and etiquette, but in the marketplace a special kind of speech was heard, almost a language of its own, quite unlike the language of Church, palace, courts and institutions (Bakhtin 1984, p.154; emphasis added).

Rabelaisian humour reflects this world of symbolic inversion, focusing on 'the bodily lower stratum', a world of carnival and grotesque humour, contrary to all existing forms of coercive social and political organization which were suspended for the time of the festivities (ibid. p.255). Bakhtin's ideas have begun to attract a wide following in contemporary cultural studies. Two specific examples will be explored here: their application to the cultural politics of 'race' and nation in postwar Britain (Gilroy 1987) and their significance in a wider discussion of the poetics and politics of transgression (Stallybrass & White 1986).

Discussing the politics of popular music from his unique position as a musician, disc jockey, and sociologist, Paul Gilroy (1987) shows how music and other popular cultural forms 'carnivalize' the dominant order of bourgeois values. Drawing on Bakhtin, he argues that music has the power 'to disperse and suspend the temporal and spatial order of the dominant culture' (ibid. p.210). Not surprisingly, then, in the tense atmosphere of the inner city, places where such music is performed become the object of police surveillance and harassment. Music venues, restaurants, and cafes have frequently been the focal point of disturbances between the police and black people. Rather like the suppression of popular culture in 19th century Britain (discussed in Ch.4), the venues of popular music have come to be regarded as a threat to social stability: 'the period allocated for recovery and reproduction is assertively and provocatively occupied instead by the pursuit of leisure and pleasure' (Gilroy 1987, p.210). This is, of course, just one instance of the 'politics of popular music' (Street 1986), other examples of which range from such deliberate interventions as the Anti-Nazi League's Rock Against Racism Campaign (Widgery 1986) to the much less self-conscious cultural politics of reggae and rap (Hebdige 1987).

Bakhtin's ideas on the carnivalization of society have also been employed in a general discussion of literary and cultural theory, focusing on the concept of transgression. Stallybrass and White (1986) refer to the way in which hierarchies of high and low extend across several related domains concerning the human body, psychic forms, geographical space, and social formation. Transgressing the rules of hierarchy and order in one domain, they argue, frequently has consequences in other domains. Attempts to suppress rowdy popular cultural forms (such as those described in Ch.4) clearly involve more than purely aesthetic objections. Crossing the boundaries of social respectability often involves an incursion into public space which is sometimes violently suppressed. The idea of transgression also helps make sense of the apparently bizarre symbolism of much popular culture where the grotesque is employed to satirize the propriety of the élite, where dominant symbols are inverted, and relations of dominance and subordination are parodied in a 'world turned upside down'. Where transgression is licensed, however, during carnivals for example, Stallybrass and White suggest that its political significance is muted. They agree with Balandier that: 'The supreme ruse of power is to allow itself to be contested *ritually* in order to consolidate itself more effectively' (ibid. 1970, p.41). Recent work, however, would suggest that this may be an over-simplification of the politics of Carnival (Jackson 1988a) which are more complex than Balandier's safety-valve theory suggests.

Stallybrass and White apply the same argument about the intersection of different domains of transgression to the geography of urban reform. In 19th-century cities, the slum, the labouring poor, the prostitute, and the sewer constituted a world in which various spheres of 'contamination' overlapped and interpenetrated. The following example, from *The Lancet* (1857), shows how the moral 'contamination' of prostitution was symbolically interwoven with references to the transgression of the boundaries between public and private space:

> The typical Pater-familias, living in a grand house near the park, sees his son allured into debauchery, dares not walk with his daughters through the streets after nightfall, and is disturbed in his night-slumbers by the drunken screams and foul oaths of prostitutes reeling home with daylight. If he look from his window he sees the pavement – his pavement – occupied by the flaunting daughters of sin, whose loud, ribald talk forces him to keep his casement closed (quoted in Stallybrass & White 1986, p.137).

The transgression of social boundaries is here represented as a transgression of spatial boundaries, cast in a language of moral outrage where the social world of debauchery, sin, and ribaldry is transposed spatially into the world of streets, parks, and pavements. This moral geography is then re-cast in metaphorical terms where a healthy body represents a healthy mind, where sanitary reform is justified in terms of moral reform, and where the social hierarchy translates into a bodily hierarchy (Fig. 7.2). There were, of course, material as well as symbolic links between the different social worlds of Victorian society: middle class families employed working-class maids, 'respectable' men resorted to female prostitutes as disgust regularly gave way to desire (see Ch.4).

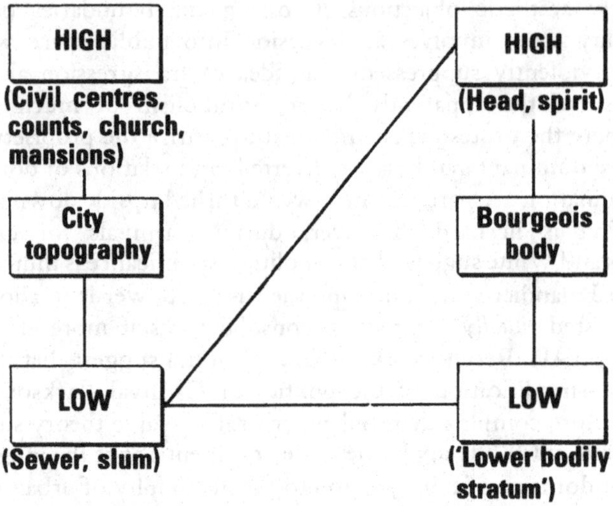

Figure 7.2 Spatial and bodily hierarchies in Victorian society

The analysis of linguistic form and symbolism provides evidence of the crucial interplay between society and space. The poetics of transgression contain a politics of language where the contradictory tendencies of conflict and compromise are, at least symbolically, resolved.

The previous examples provide an introduction to the politics of language and suggest that linguistic communities are a relatively neglected object of geographical analysis. The final example in this chapter concerns the way that geographical discourse is itself constructed and how it may disclose meaning about the politics of place. Before turning to the specific example of the European 'discovery' of Australia, some introductory comments on the analysis of discourse may be helpful.

The analysis of discourse is closely associated with the work of Michel Foucault who developed an archaeological metaphor to probe the 'discursive regularities' that unite superficially different linguistic worlds. While 'discourse' itself has a number of meanings in Foucault's work, sometimes referring to the general domain of 'things said', sometimes to an individually identifiable group of statements, and sometimes to a regulated practice that accounts for a certain number of statements (Foucault 1972, p.80), in general, the analysis of discourse is concerned with the *relation between statements* (ibid. p.31). A discourse is a 'field of regularity' that unites apparently dissimilar and unconnected statements. The analysis of discourse allows the reader to move between individual statements and the social relations of power through which those statements are articulated and given meaning, moving back and forth from 'text' to 'context'. Foucault himself applied the analysis of discourse to a very diverse range of materials, from *Madness and civilisation* (1967) to *The birth of the clinic* (1973) and *The history of sexuality* (1976). In each case, he moved from the 'level of things said' to the level of social practice, using language as the point of entry to a wider social world. The discourse of 'madness' and 'illness', for example, gave access to the cultural construction of 'sanity' and 'health' and to the institutional forms and social relations through which those categories are defined and contested. For Foucault, relations of power are never external to a discourse; rather, they are immanent in it and operate through it. These ideas inform the choice of the final example in this chapter concerning the discourse of geographical 'discovery'.

The discourse of 'discovery'

In a brilliant account of the European 'discovery' of Australia, Paul Carter shows how *The road to Botany Bay* (1987) was constructed through particular linguistic conventions that shed much light on the scholarly processes of historical research and geographical description. Subtitled 'an essay in spatial history', Carter's book argues that Australia has conventionally been seen as an empty stage on which history is acted out, a theatrical performance witnessed by an imaginary audience. Thus, Australia's leading historian, C.M.H. Clark describes the landing of the First Fleet at Botany Bay in 1788 in a language that directly recalls Virgil's description of the founding of Carthage. In Clark's account 'Some cleared ground for the different encampments; some pitched tents; some landed the stores. . .'; while Virgil wrote: 'Eagerly the Tyrians press on, some to build walls, to rear the citadel, and roll up stones by hand; some to choose the site for a dwelling and enclose it with a furrow. . .' (Carter 1987, pp.xiv-xv). Whether or not the borrowing is deliberate, it lends a heroic quality to the 'first landing' which is appropriate to the imperial cast of history to which it aspires.

The sense of theatre is even more dramatically evoked in other histories of Australia, such as the following account of the European settlement of Victoria, where the narrative is constructed as if events had been observed by an imaginary witness:

> As the sun rose on a winter's day in 1834, and the pale light successively shone on that wild coast stretching all the way from

> Bateman's Bay to the outskirts of Albany, only the sparsest signs of activity *could be seen*. Here and there the smoke drifted from a fire. On a few stretches of sand a rowing boat *might be seen*, resting well above the reach of the high tides. *An alert eye might have discerned*, in a few places, the green of a vegetable patch and the fresh unpainted wood of a hut and a new grave or two with a name and date carved on a spar or the lid of a wooden cask. Along that three thousand miles of coast, Aboriginals were probably stirring in the early morning from their sleeping places beside their tiny fires. . . (quoted in Carter 1987, p.xix; emphasis added).

The account takes the point of view of a spectator witnessing the unfolding of history, like an audience at a theatrical event. The fact that 'Aboriginals were *probably* stirring' is also indicative of the selective blindness of imperial history, a presence reluctantly admitted in accounting for an otherwise triumphant and unprecedented European 'discovery'. Even the use of place names in the preceding passage is revealing since they were, of course, only applied once the process of settlement was under way. The naming and renaming of places is a crucial aspect of geographical 'discovery', establishing proprietorial claims through linguistic association with the colonizing power. The same logic applies in every episode of 'spatial history':

> In the seventy years or so after the First Fleet's arrival, the Australian coastline was mapped . . .; the Australian interior was explored, its map-made emptiness written over, criss-crossed with explorer's tracks, gradually inhabited with a network of names; the Australian coastal strip . . . was progressively furrowed and blazed with boundaries, its estuaries and riverine flats pegged out for towns. The discoverers, explorers and settlers . . . were making spatial history. They were choosing directions, applying names, imagining goals, inhabiting the country (ibid. pp.xx-xxi).

Until its spatial history was made, Carter argues, 'Australia' only existed in the minds of the colonizers. In this sense, Australia was not 'discovered'; it had to be made. And, as Carter reveals, Australia's 'spatial history' is a cultural and not a physical product, accomplished as much linguistically as by the brute material process of settlement and colonization. Spatial history 'begins and ends in language'; by the act of naming, space is symbolically transformed into place, a space with a history (pp. xxiii-xxiv).

In the four months that Cook spent in Australian waters, he named over 100 bays, capes, and isles. Rather than impose a coherence on the process of naming by a backward-looking, imperial perspective, Carter emphasizes how the process unfolded horizontally in space and time, being articulated as a journey. He distinguishes the scientific concerns of *exploration* from the proprietorial concerns of *discovery*. Discovery rests on the assumption of a world of facts, waiting to be found; exploration is a spatial discourse in which travelling itself is knowledge, not merely the fruits of travel (ibid. p.25). Carter suggests that Cook 'took possession' of places through exploration rather than discovery: he was an explorer of horizons rather than a discoverer of countries. Exploration gives only the illusion of knowledge under the guise of naming. The names that Cook deployed obey a different, more oblique logic, the logic of metaphor (ibid. p.29). Carter's 'spatial history' demonstrates that Cook moved in a world of language as much as one of space. Geographical 'discovery' is an inherently linguistic process.

Conclusion

This chapter has suggested that a revitalized cultural geography must go beyond the mapping of languages and the geography of dialect, towards the study of language itself as the medium through which intersubjective meaning is communicated. In advocating a politics of language, differences between dialects, pidgins, creoles, and other forms of language are recognized as political rather than linguistic distinctions. Primarily, though, the development of a cultural geography that reflects the politics of language must concern itself with the spatial constitution of linguistic communities. A revitalized cultural geography will need to explore the way that language reflects and reinforces social boundaries, constituted in space and time.

Whether one is concerned with the language of 18th-century novels or with American popular culture, with 'regional' accents or the discourse of geographical 'discovery', language is a medium for communication and exchange that reflects underlying social relations of power. Like all social relations, this chapter has argued that the politics of language are constituted spatially. This theme is taken up again in the concluding chapter which reflects on the importance of geographical description in conveying the character of place and in uncovering the structures of inequality that permeate society and space.

This chapter would also suggest that geographers are now beginning to draw on a wider range of linguistic theory. The examples discussed here could be multiplied to include recent work on the media and popular culture (Burgess & Gold, 1985); analyses of the 'iconography of landscape' (Cosgrove & Daniels 1988); and studies of 'place advertising', 'decoding' the marketing strategy of public and private corporations (Burgess & Wood 1988). There is a rich field here for cultural geography the potential of which has only recently begun to be explored.

Notes

1 The theoretical position outlined here is similar to that held by the Centre for Contemporary Cultural Studies which has criticized the idealistic and ahistorical tendencies of semiological analysis for failing to recognize the effectiveness of human agency and social practice in structuring different signifying systems (Weedon *et al.* 1980). Much of the debate hinges on the extent to which linguistic signs are socially determined, rather than being purely arbitrary.

2 Lacan is significant for his emphasis on the construction of unconscious meaning in language and for his development of the idea of a gendered subject.

3 See, for example, much of the material reviewed by Zelinsky and Williams (1988) or publications such as *Geolinguistics* (from the American Society of Linguistics) and *Discussion Papers in Geolinguistics* (from North Staffordshire Polytechnic)

4 Moreover, as Driver (1985) has shown, this process is an inherently spatial one: 'A whole history remains to be written of *spaces* – which would at the same time be the history of *powers*' (Foucault 1980, p.148).

Chapter eight
An agenda for cultural geography

This book has attempted to retheorize the concept of culture and to consider some of the ways that such a retheorization might be applied in human geography. The last four chapters have provided extended examples of the application of a range of ideas from cultural studies to the analysis of popular culture, to the study of gender, sexuality, and race, and to the politics of language, focusing in each case on the *spatial constitution* of these phenomena as well as on their territorial expression. This concluding chapter attempts to draw together some of the more general threads of the argument and to consider an appropriate agenda for future research in cultural geography.

Specifically, this chapter provides an overview of two current theoretical debates in which cultural studies are central to the development of a reconstituted human geography. The first concerns the emergence of a range of interpretive approaches to the study of culture and society; the second concerns the concept of post-modernism, recently extended from art and architecture into geography and the social sciences (Dear 1988, Soja 1989). This chapter, like the rest of the book, traces the relationship between culture and society, emphasizing the political dimensions of this relationship implicit in the idea of cultural politics. This chapter also attempts to revise conventional definitions of culture and landscape by emphasizing a *plurality of landscapes*, reflecting a *plurality of cultures* or 'ways of seeing' (Berger 1972). This leads, finally, to an attempted reformulation of the concepts of 'culture' and 'geography' that have been central to the whole book.

Culture and interpretation

The principal alternative to the inclusive view of culture proposed by Tylor (cf. Ch.2) is the idea of *culture as interpretation*. Interpretative approaches to culture are closely associated with the work of Clifford Geertz, an anthropologist whose work has attracted a wide following beyond the small group of experts who share his professional interest in the anthropology of South East Asia and Morocco. Geertz defines anthropology as an interpretative science in search of meaning, not an experimental one in search of laws (Geertz 1973, p.5). His work stresses the way that social relations involve a continuous process of interpretation and reinterpretation. Society is not static or fixed but dynamic and fluid, made up of a constant flux of individual practices, patterned according to social rules from which structures may be inferred. According to Geertz, the anthropologist's concern in doing fieldwork is to produce an account of another society by stepping into this flow of events and constructing an interpretation that can be re-presented to another audience. Indeed, ethnography takes its name from the *inscription* of social action: from writing it down.

Significantly, 'ethnography' itself contains a duality of meaning. In one sense, ethnography refers to the process of gathering anthropological data (through fieldwork, the collection of genealogies etc.). In another sense, it refers to the end product of that process (the production of a written account or ethnographic text). The double meaning is symptomatic of the anthropologist's ambiguous relationship with his or her informants. Anthropologists provide an interpretation of other people's actions for a professional audience. But people's actions are themselves based upon a continuous process of interpreting and reinterpreting the actions of others, to which the anthropologist supplies an additional layer of meaning. As Geertz expresses it: 'what we call our data are really our own constructions of other people's constructions of what they and their compatriots are up to' (Geertz 1973, p.9). This idea encompasses what is sometimes called a 'double hermeneutic', involving the interpretation of an interpretation. It is a complex idea that is worth exploring in more detail.

The field of hermeneutics developed initially in relation to the interpretation of Biblical texts where theologians offered alternative readings of the scriptures. The term was then extended to refer to any process of interpretation including the process of understanding society itself. Several authors, including Geertz, likened the understanding of society to the interpretation of a written text: 'Arguments, melodies, formulas, maps, and pictures are not idealities to be stared at but texts to be read; so are rituals, palaces, technologies, and social formations' (Geertz 1980, p.135). Society can be likened to a text, according to

Geertz, in the sense that any social practice (even the wink of an eye) is capable of multiple interpretations. All ethnographic accounts are therefore 'partial truths' not just in the sense of being incomplete but also because they are written from a particular point of view (cf. Clifford 1986). Ethnography, then, as Geertz conceives it, consists not merely of 'plaguing subtle people with obtuse questions' (op. cit. 1973 p.29) but also of producing written texts which offer an interpretation of other people's interpretations of events they have themselves experienced. Thus Geertz' insightful analysis of cockfighting in Bali describes 'a Balinese reading of Balinese experience, a story they tell themselves about themselves' (ibid. 1973, p.448)

Many social scientists have been stimulated by Geertz' formulation of an interpretative anthropology and the idea of society-as-text is now quite commonplace in other disciplines, ranging from literature to history (cf. LaCapra 1983, Darnton 1984). Geographers have also begun to 'read' the landscape, to refer to its 'biography' and to employ the metaphor of landscape-as-text (Meinig 1979, Ley 1987, 1988a). Interpretative approaches have attracted a wide following. But they have also been treated to hostile criticism. Paul Shankman, for example, asks rhetorically whether:

> A movement without direction, a program troubled by inconsistency, an approach that claims superiority over conventional social science but is limited by the absence of criteria for evaluating alternative theories, and type cases that do not necessarily support the interpretive theory – can this be the basis for a different anthropology and a major intellectual movement? (Shankman, 1984, p.270).

He describes Geertz' interpretative approach as a style, a fashion, or a genre that is alluring, exciting and even glamorous, but one that is ultimately guilty of a sterile elegance.

The most serious criticism of interpretative approaches such as Geertz', however, is their tendency to advocate an interactionist view of society which shows too little concern for the structural constraints on individual action. Geertz is himself aware of this criticism. His study of the theatre state in 19th-century Bali (Geertz 1980), for example, gives sustained attention to the political dimensions of social life. In this case, Geertz approaches politics as a domain of social action defined in terms of people's control over resources. Within social geography, the question has been posed by David Ley in terms of 'projecting local circumstances into a broader discourse' (1988b, p.134). The relationship between culture, politics, and society that this question raises is developed further in the following section.

Sheep raids and revolutions: culture, politics, and society

In a memorable passage from his essay on ethnography as 'thick description', Geertz describes the way that 'social actions are comments on more than themselves': 'small facts speak to large issues, winks to epistemology, or sheep raids to revolutions, *because they are made to*' (Geertz 1973, p.23; emphasis added). It is, in other words, up to the ethnographer to make these connections through the construction of an ethnographic text. As Clifford (1986 p.6) teasingly suggests, ethnographies are works of fiction, in the sense of 'something made or fashioned'. He insists, though, that they are 'true fictions' rather than total fabrications.

This concern with the production of ethnography-as-text has led to a rash of experiments in ethnographic writing including those by Geertz himself, exploring different textual strategies for 'writing cultures' (cf. Clifford & Marcus 1986). In *Negara* (Geertz 1980), for example, Geertz separates the 'narrative' from the 'scholarly apparatus' of notes and references, devoting almost an equal amount of space to each. He is above all, however, a deft exponent of the essay form which he uses to 'exoticise the familiar', applying anthropological methods to studies of the American legal system or the practice of literary criticism which, he reveals, no less than any 'primitive' society, all possess their own forms of 'local knowledge' (Geertz 1983). What Geertz is less adept at showing, however, is how some versions of knowledge have greater power than others: how scientific knowledge can be used to legitimize racist ideologies, for example, or how religion and professional medicine have marginalized other forms of knowledge, such as magic and witchcraft, achieving a privileged position, sanctioned by the state and royal patronage (cf. Thomas 1971). In other words, some forms of knowledge acquire hegemonic status where their preferred readings are accepted as 'natural' or 'common sense'.

Even in his essay on ideology as a cultural system, however, Geertz (1973) does not approach ideology in the critical sense of an 'unexamined discourse' the meanings of which conceal material interests. Geertz wishes in fact to 'defuse' the concept of ideology by finding a 'non-evaluative' alternative to its present pejorative implication. He rejects an interest-based theory of ideology as psychologically 'anemic' and sociologically 'too muscular' (ibid. p.202). Such views, he says, ignore the complex *motivations* of those who subscribe to a particular ideology. One can, however, avoid this dilemma by judging an ideology according to the pragmatists' maxim, in terms of its social effects or *consequences*, irrespective of individual motivation or intention.

If Geertz' work is weak in its approach to politics and ideology, he is much more convincing when dealing with the structure and formation of symbolic systems. While one can accept that material frustrations and contradictions are often resolved symbolically, through 'rituals of resistance' (Hall & Jefferson 1976), it is much harder to specify exactly how particular symbols come to express those frustrations: 'how the trick is really done' as Geertz describes it (1973 p.207). Some anthropologists, such as Lévi-Strauss (1966), discuss the process in terms of *bricolage* (or 'do-it-yourself'), where symbols are chosen because they are *bonne à penser* (literally 'good to think'). Others speak of an almost infinite 'forest of symbols' (Turner 1967) within which certain associations, such as those concerned with the human body, are sufficiently common as to appear almost 'natural' (Douglas 1970).

Geertz (1973, p.213n) offers a promising alternative to the tautologies of 'natural' or universal symbols without entirely accepting the 'arbitrariness of the sign'. He argues that ideology is a *cultural system*, an intricate structure of interrelated meanings. This multiplicity of meanings, he argues, is the key to understanding symbolic systems. For much of the power of symbolism, as with metaphor (Lakoff & Johnson 1980), derives from its ambiguity. Rather than reducing complex social meanings to one literal truth, symbolic forms allow more subtle, multiple meanings to coexist. Ideology, in Geertz' terms then, is not simply a *distortion of meaning*; it is a recasting of meaning at one level of significance at another level. The co-existence of dual or multiple representations (and how to arbitrate between them) is one of the key problems confronted by 'post-modernism' and its critics.

Poetics and politics: the dilemmas of post-modernism

The meaning of 'post-modernism' is shrouded in uncertainty. A recent dictionary entry declares: 'This word has no meaning. Use it as often as possible' (quoted in Featherstone 1988, p.195). Tracing the route by which it entered human geography, from art and architecture, offers therefore a useful starting point. Post-modern architecture is distinguished by its combination of styles, by its preference for the vernacular over the academic, and by its rejection of strict historicism. Post-modernism tends towards the hybrid, with a penchant for parody and kitsch, and a determined irreverence towards the canons of good taste.

But the language of post-modernism (Jencks 1981) can be quite subtle, often distinguished by a *dual coding* of meaning (the Chippendale pediment on a New York high-rise is a famous example, with the language of 18th century furniture superimposed on the language of contemporary architecture). Post-modern buildings can therefore be read on at least two levels, addressing a knowledgeable group of architects and historians, for example, trained to read stylistic cross-references and quotations, while simultaneously appealing to the public at large as a form of visual entertainment. Post-modern design is anti-élitist, aiming to appeal to a range of tastes including the vernacular. It employs a complex language, with heavy metaphoric content, drawing on previous styles and subverting architectural conventions by self-conscious use of allusion, visual parody, and wit.

The language of post-modernism has been adopted by social scientists in an attempt to highlight the way that knowledge is constructed and contested rather than being simply received. The 'post-modernization' of geography (Soja 1987) has signalled a renewed interest in the problems of geographical description (Daniels 1985, Lewis 1985) and encouraged an enhanced sensitivity towards textual strategy (Gregory 1988). Many of these developments are paralleled in anthropology and the other human sciences, all of which, according to Marcus and Fischer (1986), are facing a period of crisis and experiment. The lack of self-consciousness about modes of representation that Geertz once noted (1973 p.19n) has been replaced by an almost obsessive degree of self-reflection. Taking post-modern fiction as its model, the shackles of a naïve empiricism have been thrown off and social scientists are eagerly experimenting with a wide variety of modes of representation. The narrative self-consciousness of post-modern authors like Garcia Marquez and Borges are beginning to appear in academic social science, mirrored in geography by the linguistic experiments of Gunnar Olsson and Allan Pred (Olsson 1980, Pred 1984). Despite the avowed concern of post-modernism with both politics and poetics however, the latter has, with few exceptions, received much more emphasis than the former (cf. Gregory 1987).

The challenge of post-modernism can be represented as the problem of linking aesthetic and intellectual trends to changes in the material world, without simply reducing post-modernism to 'the cultural logic of late capitalism' (Jameson 1984). For some geographers, post-modernism is nothing more than 'the cultural clothing of flexible accumulation' (Harvey 1987b, p.279). For others, however, the concept of flexible or specialized accumulation is itself of dubious value: 'overly flexible and insufficiently specialised', according to one recent critic (Sayer 1988). The 'post-modern condition' (Lyotard 1984) clearly has a sociology and a geography that extend well beyond the superficially 'cultural'. But there have as yet been relatively few attempts to specify the relationship between post-modern culture and the changing contours of contemporary capitalism (Lash & Urry 1987, provide one of the exceptions, centred on the notion of 'disorganized capitalism'). Even the basic steps of defining the process of post-modernization or the nature of post-modernity have so far produced little consensus (cf. Featherstone 1988). One of the few areas of consensus within the whole post-modernism debate concerns the inadequacy of a unitary concept of culture. For geographers, of course, a plurality of cultures also implies a multiplicity of landscapes.

Cultures and landscapes

This book has consistently rejected a unitary and élitist view of culture. It has focused instead on the plurality of cultural forms through which dominant meanings are contested. For if, as Cosgrove (1985a) insists, landscape is 'a way of seeing', then there are potentially as many ways of seeing as there are eyes to see. Other social scientists have begun to recognize the implications of these ideas and to employ a variety of landscape metaphors in their work. Richard Hoggart's 'landscape with figures' (1957) was among the first metaphorical uses of this kind. Christine Stansell's 'sexual landscapes' and 'moral geography' (1986) is a more recent example. But other authors, describing the cultural politics of particular times and places, have also resorted to spatial analogies such as 'maps of the historical landscape' (Schorske 1980), or the 'mire of the macadam' and the 'modernism in the streets' (Berman 1982).

Reading 'the iconography of landscape', through art and architecture, cartography and design, represents one of the most prominent new directions in cultural geography (Cosgrove & Daniels 1988), arguing from a world of exterior surfaces and appearances to an inner world of meaning and experience. While some have employed this geographical vocabulary in a purely figurative sense, theorizing the relationship between culture and society in very diverse ways, all would agree that contemporary culture is comprised of multiple strands. A reconstituted cultural geography must therefore be prepared to examine the multiplicity of landscapes that these plural conceptions of culture inform. From the range of studies that exemplify some of these new directions, two have been selected for further consideration: Steven Kern's analysis of the changing consciousness of time and space at the turn of the 19th century (Kern 1983), and Carl Schorske's account of culture and politics in *fin de siècle* Vienna (Schorske 1980).

In *The culture of time and space, 1890-1918*, Steven Kern (1983) examines a period of intense technological change in North America and Western Europe, from the invention of the telephone (in 1876) through the development of the wireless telegraph, X-ray, and cinema, to the invention of the bicycle, the automobile, and the aeroplane.

Drawing on contemporary developments in psychiatry and phenomenology, Kern relates these technological changes to various cultural and academic developments such as the invention of the stream-of-consciousness novel, Freudian psycho-analysis, Cubism, and relativity theory. Proceeding by analogy and by the identification of 'compelling similarities', Kern makes a number of daring associations between the changing experience of time and place, and the various diplomatic and military crises of the period leading up to World War I. For example, he notes the connection between Cubism and camouflage which Picasso mentioned in a letter to Gertrude Stein, tracking the association down to the inventor of camouflage who himself explicitly acknowledged the inspiration of Cubism.

During these years of social and economic turmoil, Kern suggests, a plurality of times and spaces was affirmed. He notes the levelling of hierarchies throughout Western culture (from the 'democratic' architecture of Louis Sullivan to the negation of the distinction between subject and background in Cubist painting), noting parallels with the levelling of aristocratic society, the rise of political democracy, and the blurring of the boundary between sacred and profane space in contemporary religion. The broad sweep of Kern's argument can be compared with Schorske's study of *fin de siècle* Vienna which examines a similar period and an equally wide range of examples (from art and architecture to psychiatry and music), but on the more limited canvas of a single city [1].

When the Liberals came to power in Austria in 1860, they proceeded to transform the institutions of the state, replacing an aristocratic absolutism with the principles of constitutionality. Urban redevelopment in Vienna came to symbolize these changing power relations. For the Liberals, Vienna represented 'their political bastion, their economic capital, and the radiating centre of their intellectual life' and 'the projection of values into space and stone' (Schorske 1980, pp.24–5). The Liberals transformed the city in their own image, a process that had been virtually completed by the turn of the century when a landslide victory by the Christian Social Party swept them from power.

While Hausmann was transforming Paris, monumentalizing the Second Empire (Harvey 1985a, Woolf 1988), the reconstruction of Vienna centred on the Ringstrasse which became the symbol of the age: 'an iconographic index to the mind of ascendant Austrian liberalism' (Schorske 1980, p.27). Although the Liberals undertook a number of less dramatic projects, such as the channelling of the Danube, the establishment of a public health system, and the construction of new parks and hospitals, it was the peripheral boulevard and its monumental buildings that captured the public imagination and impressed by their sheer scale. The new Parliament, Rathaus, University, and Burgtheater symbolized the triumph of constitutional rule and the rise of an alternative secular culture, replacing the imperial palace, the Gothic cathedral, and the military garrison.

The strength of Schorske's analysis is its blending of culture and politics.[2] Tastes are not simply inscribed in the landscape. Rather, the political and economic forces that lay behind the new construction are traced in detail. The Ringstrasse itself, to take just one example, was built under the auspices of a City Expansion Committee which used the proceeds from the sale of military land to finance a City Expansion Fund. Apart from demonstrating these close material links between past and present, private and public, the visionary and the pragmatic, Schorske also demonstrates how the redevelopment of the Ringstrasse was actively contested by rival visions of the new urban order becoming 'the anvil against which two pioneers of modern thought about the city, Camillo Sitte and Otto Wagner, hammered out ideas of urban life and form' (ibid. p.25).

For a humanitarian like Sitte, the Ringstrasse embodied the worst features of a heartless, utilitarian rationalism. His vision of the future drew on the ideals of John Ruskin and William Morris, with architecture and people in communitarian union. In contrast, Wagner's vision of city planning aimed to foster rational economic growth by improving the efficiency of urban transportation. His designs sought to embody 'the colossal technical and scientific achievements . . . of modern mankind' (ibid. p.74). He was a utilitarian and a functionalist, dedicated to 'aesthetic engineering' and more in tune with art nouveau than with the arts and crafts movements. The battle between Sitte and Wagner represents the opposition between two conflicting ideological visions and personifies the plurality of landscapes that are possible within a single city at a particular historical moment. It is the kind of tension that cultural geographers like Denis Cosgrove have tried to capture in describing the relationship between symbolic landscape and social formation (Cosgrove 1985b) or which David Ley has tried to emulate in his studies of the changing landscapes of inner Vancouver (Ley 1987).[3]

Before concluding this review of the potential of a reconstituted cultural geography, it is worth returning to the central concepts of 'culture' and 'geography'. In stark contrast to much of the existing literature in cultural geography, it could be said that theoretical debate in contemporary cultural studies has tended to outstrip empirical research. These concluding remarks, which attempt to identify some of the most significant insights from contemporary cultural studies and to situate them in terms of current debates in human geography, therefore have a rather programmatic flavour. But, it is suggested, the agenda for cultural geography will be shaped in relation to these trends.

Culture . . .

This book has tended to concentrate on the 'cultural' in an adjectival sense rather than on 'culture' itself in more substantive terms. Following Raymond Williams (1958), the cultural has been defined in relation to 'real material forces' and the social relations that those forces evoke. The stuff of culture is more elusive, best approached obliquely in terms of the *processes* through which meanings are constructed, negotiated, and experienced. This apparent reticence to approach culture 'head on' reflects the problems that are encountered by more 'concrete' definitions. Once the discussion moves away from the adjectival sense of culture, analysis quickly degenerates into the reifications of 'cultural baggage', 'culture contact', or the 'clash of cultures', criticized in Chapter 2 as characteristic of the worst kind of culturalism. Too often in the past, cultural geography has degenerated into an inventory of 'culture traits', mapping physical artifacts like barns and fences that are visible in the landscape. Though there have been some welcome developments in this field, including urban as well as rural artifacts, and popular as well as élite cultural traits (Rooney *et al.*, 1982), this kind of approach still takes insufficient notice of recent developments in cultural theory. If cultural geography is to be revitalized, to paraphrase Stedman Jones (1983), 'it cannot be by the defensive reiteration of well tried and by now well worn formulae. It can only be by an engagement with the contemporary intellectual terrain – not to counter a threat, but to discover an opportunity' (ibid. p.24).

The alternative approach adopted here is to view culture as the medium or idiom through which meanings are expressed. If one accepts the preceding arguments for a plurality of cultures, then 'culture' is the domain in which these meanings are contested. Starting from this point, cultural studies has tended to focus on subordinate groups and this book has been no exception. But studies of women and minority groups, for example, must now be superseded by studies of gender and racism which focus on the relations between dominant and subordinate groups. Similarly, cultural geographers should not ignore dominant cultures where a whole new agenda remains to be investigated. Recent work on nationalism, monarchy, and heritage, for example (reviewed by Thrift 1989), suggests a range of avenues for future research, whether or not this involves a conception of 'class cultures' as Thrift himself implies. Working with dominant cultures entails its own political and ethical problems, no less than more traditional work with 'minority' groups. But, like the questions posed by the 'repatriation' of anthropology, returning from overseas to the domestic world of advanced industrial societies (Burgess 1984), there is the potential here for a radical transformation of the whole field of cultural studies.

As these developments suggest, cultural theory is not a closed book. Many areas of debate are still actively contested and there is ample scope for spatially sensitive analysis. Some areas in need of particularly urgent clarification include the extent to which cultural strategies are consciously premeditated; the degree to which individual spontaneity and creativity are constrained by structure and circumstance; the tensions between élite and popular culture; the scope for active resistance in the process of consumption; and the development of a critical but non-reductionist conception of ideology. These issues, among others, also call out for further empirical research, notwithstanding the greater theoretical sophistication now available as a result of geography's encounter with contemporary social theory. This chapter therefore concludes with a consideration of the immediate disciplinary context of cultural geography and with the perennial question of whether, why, and how much, 'geography matters' (Massey & Allen 1984). There are grounds for considerable optimism here as human geography begins to take a more central place within the social sciences.[4] Indeed, geographers are now better placed than they have been for many years to contribute to social and cultural theory, rather than simply drawing parasitically from it.

. . . and geography

This book has explored a variety of approaches to the geography of culture other than those that focus exclusively on landscape. Even within the landscape tradition, it has emphasized the idea of landscape as a social construction or a 'way of seeing' rather than as reducible to a series of physical traits. Among alternative approaches to the landscape and 'man-environment' traditions, this concluding section will consider three possibilities which seem best suited to a reconstituted cultural geography: the theory of uneven development; the concept of spatial divisions of labour; and the reciprocal links between social relations and spatial structures.

The theory of uneven development has been expounded within geography by a number of Marxist writers, of whom Neil Smith has been the most prominent. Smith describes uneven development as 'the systematic geographical expression of the contradictions inherent in the very constitution and structure of capital' (1984, p.xi). Uneven development at urban, national, and global levels results from the tension between two inherently divergent tendencies within capitalism:

the spatial fixity of capital, and the need for capital mobility in order to offset the law of diminishing profits. This tension results in permanent contradictions and periodic crises involving the restructuring of geographical space (from gentrification and counter-urbanization, through deindustrialization and regional decline, to imperialism, geopolitics, and war).

Smith's theory of uneven development may not, at first sight, seem a likely choice as an appropriate theory with which to approach the geographical study of culture. His emphasis on the 'inner necessity' and 'economic logic' of capitalism suggests a thoroughly de-cultured view of society where social relations are rigidly structured by an inflexible political economy, and where the scope for human agency is severely constrained as a disembodied capital 'stalks the earth' (Smith 1984, p.49). But it is also possible to see in Smith's work an attempt to grapple with the fundamentally social character of human existence, with the social construction of Nature, and with the variety of ways in which capital accumulation is mediated geographically. Coupled with Harvey's recent work on the urbanization of capital and its implications for human consciousness (Harvey 1985a, 1985b), it is possible to see how a theory of uneven development might form the basis for a more rigorous cultural materialism. Indeed, Smith's work on gentrification begins to address these questions, including the more intangible aspects of 'yuppie' culture as well as the 'harder' lineaments of political economy (Smith 1987b). Smith raises questions about the cultural and ideological implications of describing gentrification according to 'frontier' and 'wilderness' analogies, probing the relationship between social restructuring and the 'American Dream'. He develops Sharon Zukin's conception of gentrification as the intersection of culture and capital at the urban core (Zukin 1986), extending these ideas in directions hitherto unexplored with the exception of a handful of journalists such as Patrick Wright (1985). But with *science parks* replacing *industrial estates*, there is clearly a moral and cultural geography behind urban reinvestment and economic restructuring, as well as a purely rational, economic one.

Doreen Massey would also accept that economic restructuring involves cultural as well as political and economic processes. Her discussion of the North-South divide deals not only with changing investment patterns, employment and income differentials, and geographic mobility, but also with the changing imagery of North and South. In particular, she notes that the image of the North has been reworked of late:

> There is less mention of satanic mills. More, the talk (in the south) is of how *wonderful* the countryside is, and the quality of life it is

> possible to have, and of how low house prices are. The recent study, *Northern Lights*, picked out the top ten northern towns . . . using criteria such as *Good Food Guide* restaurants, *Michelin Guide* hotels, number of antique shops, presence of golf clubs [as well as] proportion of social classes 1 and 2 and the number of households with more than one car (Massey 1988, p.17).

Images of North and South, like those of country and city, contain a wealth of contradictory messages that need decoding as much in terms of cultural as of economic geography.

In her *Spatial divisions of labour* (1984), Massey suggests that the impact of industrial restructuring on particular localities has been highly uneven, reflecting previous rounds of capital investment and the evolution of distinctive regional economies. Focusing as she does on the geography of production, it is not surprising that economic issues are more prominent than cultural ones. However, it is possible to extend Massey's argument about spatial divisions in the geography of production to the constitution of society in general. Massey suggests as much herself, describing her overall aim as to trace the reciprocal links between the geography of industry and the underlying structures of society. As recent locality studies suggest, this relationship is a reciprocal one in the sense that particular places are not just the passive recipients of structural change; they also have a 'proactive capacity' to influence those changes, to resist and redefine them (cf. Cooke 1988).

Although Massey's work gives relatively little direct consideration to cultural issues, it contains a number of tantalizing references which hint at the potential relevance of a 'spatial divisions of labour' approach for cultural analysis. Her discussions of regionalism, of the social definition of skill, and the gender division of labour are all cases in point. Casual references to the 'individualistic stroppiness of Merseyside workers' and to the 'organised discipline of miners from South Wales' (Massey 1984, p.58) call for more elaboration. The analysis of 'local political cultures' is part of a more general revival of regional geography, the contours of which are only now becoming apparent (Gilbert 1988, Pudup 1988, Sayer 1989).

The locality studies initiative funded by the UK Economic and Social Research Council (Cooke 1986), to which Massey's work gave rise, has already begun to address some of these issues, but they have only scratched the surface of some very complex problems. Understanding 'local culture' cannot be reduced to a selective emphasis on certain arbitrary aspects of a region's rural or industrial 'heritage' for the purpose of promoting tourism, as some recent work appears to suggest.[5] Much larger questions are involved concerning the relationship between local and national cultures, urban and rural environments, concepts of work and leisure, ideas about the past and prospects for the future.

One further area in which Massey's ideas need urgently to be extended concerns the geography of 'race' and racism (cf. Jackson 1987). While she provides a preliminary survey of the variable significance of class and gender relations in different localities, there is virtually no discussion of race. This may be because she feels that racism is not the same kind of structure as capitalism or patriarchy, with its own autonomous powers and independent dynamic. But further clarification is obviously needed. Any agenda for cultural geography should therefore include an exploration of how particular 'racial divisions of labour' are constituted geographically and historically, how labour is selectively 'racialized', and how specific intersections of race, class, and gender are played out across space and over time.

Massey's work serves as an important bridge between theories of uneven development and ideas about the spatial constitution of society. For, contrary to the aspatial theorizing of some sociologists and economists, the world does not take place on the head of a pin (Massey 1985, p.51); it is geographically and historically variable. Moreover, as Massey insists, the relationship between society and space is a *reciprocal* one; not a one-way street whereby spatial patterns passively reflect social processes. The impact of political and economic change, in Massey's formulation, is mediated by the effects of previous changes which, to employ one of her geological metaphors, become sedimented over time in local and regional cultures.

One of the most exciting theoretical challenges for a reconstituted cultural geography is to incorporate the insights of the 'society and space' debate (Gregory & Urry 1985) into cultural studies, emphasizing the extent to which *the social is spatially constituted*:

> The fact that processes take place over space, the facts of distance or closeness, of geographical variation between areas, of the individual character and meaning of specific places and regions – all these are essential to the operation of social processes themselves. . . Nor do any of these processes operate in an environmentally characterless, neutral and undifferentiated world. Geography in the fuller sense implies not only spatial distance but also physical differentiation, of terrain, of vegetation, of climate (Massey 1985, p.52).

Lest this be seen as taking geography back to a previous era of 'areal differentiation' and environmental determinism, it is important to consider what these remarks might imply for a fully reworked cultural geography. Cultural geographers should be concerned not just with tracing the effects of successive rounds of capital investment and disinvestment in particular regions and localities, accounting for those differences in terms of their distinctive histories and geographies. They should also begin to explore the diverse ways in which those processes are *culturally encoded*: how working-class history is appropriated and symbolically transformed in the course of urban redevelopment, for example. These processes might be theorized in terms of the concept of 'cultural capital' (Bourdieu & Passeron 1977, Bourdieu 1984). But so far these ideas have found few adherents in human geography.

Studying the spatial constitution of society through the mediating effects of culture, a revitalized cultural geography must avoid becoming a rigid geometry of 'spatial relations', while at the same time eschewing the more indulgent and idiosyncratic aspects of the humanistic study of 'sense of place'. The challenge that this book issues is to open up the field of cultural geography to a range of theoretical perspectives that stress the social and political construction of culture, without surrendering the specific insights of human geography. Sketching a broad agenda for cultural geography should not, however, be taken as an invitation to undisciplined inquiry. Rather, cultural geographers should be encouraged to draw on the methodological rigour of neighbouring disciplines, such as anthropology (with its inherently comparative method) and history (with its passion for hard data and verifiable evidence).

While the value of some post-modernist forms of inquiry has been recognized, the danger of reducing culture to a political struggle over language should not be ignored (Chambers 1986). It is an impoverished view of culture that stresses text, sign, or discourse to the exclusion of context, action, and structure. Meanings must always be related to the material world from which they derive. For, while cultures contain a multiplicity of meanings, it is always possible to arbitrate among them by means of the interests they represent. Dominant cultures are not the same as subordinate ones, neither is popular culture the equivalent of élite culture: they vary along a scale of cultural power (Clarke *et al.* 1976, p.11). Indeed, cultures are *collective* representations the communicative value of which depends on their shared or social meaning. Local cultures, as Suttles (1984) maintains, have a cumulative texture, albeit one that is fluid and contested. But a revitalized cultural geography would be just as interested in seeing how dominant values are institutionalized through the operation of hegemonic forces at the national level as it would be in tracing the detailed contours of particular subcultures at the local level.

This book has explored some of the diverse 'maps of meaning' through which groups and individuals make sense of their social world.

Like any cartographic image, 'maps of meaning' codify knowledge and represent it symbolically. But, like other maps, they are ideological instruments in the sense that they project a preferred reading of the material world, with prevailing social relations mirrored in the depiction of physical space. Some meanings are dominant; others result from struggle against the dominant order. As with every map, however, a certain ambiguity always remains. Cultural maps are capable of multiple readings. But, as this book has tried to demonstrate, dominant readings never go completely unchallenged; resistance is always possible.

Notes

1 In this respect, Schorske's work is most usefully compared with Janik & Toulmin's study of *Wittgenstein's Vienna* (1973).
2 Schorske goes on to analyze Freud's *Interpretation of dreams*, the paintings of Gustav Klimt, the image of the garden in Austrian literature, and the art and music of Kokoschka and Schoenberg, against a background of political and social change in Vienna, represented by 'the birth of urban modernism' in the Ringstrasse.
3 In his more recent work on world fairs and international expositions, Ley draws on the notion of a 'society of the spectacle' (Debord 1983) to describe the 'landscape of heroic consumption' (Ley 1988a).
4 As one recent commentator confidently declares: 'geography can no longer be seen as the Cinderella subject of the social sciences, borrowing most of its methods and ideas from other disciplines. Instead it is able to make its own claims that social, economic and political processes cannot be discussed without being informed by geographical analysis' (Cochrane 1987, p.354). By contrast, Andrew Sayer's assessment of 'the difference that space makes' is more cautious: 'social theory has managed to pay space scant attention without too much trouble', while 'those theorists who have been preoccupied with space have not been able to say very much about it' (Sayer 1985, p.65).
5 Compare John Urry's recent work on the decline of the traditional seaside resort which he relates to the absence, in such places, of well-preserved historical sites which can be 'sacralized, packaged and viewed' for the (post-modern) tourist. Some of the least prepossessing industrial sites, he argues, have been transformed into the most successful tourist ventures: 'the worse the "industrial" experience, the more authentic the resulting attraction seems to be' (Urry 1988, p.50).

Bibliography

Agnew, J. A. & J. S. Duncan, 1981. The transfer of ideas into Anglo-American human geography. *Progress in Human Geography* **5**, 42–57.

Aitchison, J. W. & H. Carter, 1987. The Welsh language in Cardiff: a quiet revolution. *Transactions, Institute of British Geographers* New series **12**, 482–92.

Althusser, L. 1969. *For Marx*, Harmondsworth: Penguin.

Altman, D. 1986. *AIDS and the new puritanism*, London: Pluto Press.

Amis, M. 1986. Double jeopardy: making sense of AIDS. In *The moronic inferno and other visits to America*, 187–98. London: Jonathan Cape.

Anderson, K. J. 1987. The idea of Chinatown: the power of place and institutional practice in the making of a racial category. *Annals, Association of American Geographers* **77**, 580–98.

Anderson, K. J. 1988. Cultural hegemony and the race-definition process in Chinatown, Vancouver: 1880–1980. *Society and Space* **6**, 127–49.

Anderson, N. 1923. *The hobo*, Chicago: University of Chicago Press.

Arnold, M. 1869. *Culture and anarchy: an essay in social and political criticism*. London: Smith, Elder.

Bachelard, G. 1964. *The poetics of space*. New York: Orion Press.

Bailey, P. 1978. *Leisure and class in Victorian England: rational recreation and the contest for control, 1830–1885*. London: Routledge & Kegan Paul.

Bailey, P. (ed.) 1986. *Music hall: the business of pleasure*. Milton Keynes: Open University Press.

Bailey, P. 1987. Leisure, culture and the historian: confessions of a vulgar culturalist. Introduction to the paperback edition of *Leisure and class in Victorian England*, 8–20. London: Methuen.

Bakhtin, M. 1984. *Rabelais and his world*. Bloomington, Indiana: Indiana University Press.

Balandier, G. 1970. *Political anthropology*. Harmondsworth: Penguin.

Barnes, J. A. 1979. *Who should know what? Social science, privacy and ethics*. Harmondsworth: Penguin.

Barrell, J. 1980. *The dark side of the landscape: the rural poor in English painting, 1631–1741*. Cambridge: Cambridge University Press.

Barrell, J. 1982. Geographies of Hardy's Wessex. *Journal of Historical Geography* **8**, 347–61.

Barrett, M. 1980. *Women's oppression today: problems in Marxist feminist analysis*, London: Verso.

Becker, H. S. 1963. *Outsiders.* New York: Free Press.

Bell, C. & Bell, R. 1972. *City Fathers.* London: Penguin.

Benedict, R. 1935. *Patterns of culture.* London: George Routledge & Sons.

Berger, J. 1972. *Ways of seeing,* Harmondsworth: Penguin.

Berman, M. 1982. *All that is solid melts into air: the experience of modernity.* New York: Simon & Schuster.

Berry, B. J. L. 1980. Inner city futures: an American dilemma revisited. *Transactions, Institute of British Geographers* New series **5**, 1–28.

Bhavnani, K. K. & R. Bhavnani, 1985. Racism and reaction in Britain. In *A socialist anatomy of Britain,* D. Coates, G. Johnston & R. Bush, (eds.), 146–59. Cambridge: Polity Press.

Billinge, M. 1984. Hegemony, class and power in late Georgian and early Victorian England: towards a cultural geography. In *Explorations in historical geography: interpretative essays,* A. R. H. Baker, & D. Gregory, (eds.), 28–67. Cambridge: Cambridge University Press.

Birch, B. P. 1981. Wessex, Hardy and the nature novelists. *Transactions, Institute of British Geographers* New Series. **6**, 348–58.

Blackburn, R. 1988. Raymond Williams and the politics of a New Left. *New Left Review* **168**, 12–22.

Boal, F. W. 1969. Territoriality on the Shankill-Falls divide, Belfast. *Irish Geography* **6**, 30–50.

Boal, F. W., M. A. Poole, R. Murray, & S. J. Kennedy, 1977. *Religious residential segregation and residential decision making in the Belfast urban area.* SSRC Project Report No. HR1165.

Bolt, C. 1971. *Victorian attitudes to race.* London: Routledge & Kegan Paul.

Booth, C. 1889. *Life and labour of the people.* London: Williams & Norgate.

Booth, C. 1890. *In darkest England and the way out.* London: Salvation Army International Headquarters.

Bourdieu, P. 1977. *Outline of a theory of practice,* Cambridge: Cambridge University Press.

Bourdieu, P. 1984. *Distinction: a social critique of the judgement of taste.* London: Routledge & Kegan Paul.

Bourdieu, P. & J.-C. Passeron, 1977. *Reproduction in education, society and culture.* London: Sage.

Bourne, J. & A. Sivanandan, 1980. Cheerleaders and ombudsmen: the sociology of race relations in Britain. *Race and Class* **21**, 331–52.

Bowen, M. 1981. *Empiricism and geographic thought: from Francis Bacon to Alexander von Humboldt.* Cambridge: Cambridge University Press.

Bowlby, S., J. Foord, & L. McDowell, 1986. The place of gender relations in locality studies, *Area* **18**, 327–31.

Boyer, P. 1978. *Urban masses and moral order in America, 1820–1920.* Cambridge, Mass.: Harvard University Press.

Bratton, J. S. (ed.) 1986. *Music hall: performance and style.* Milton Keynes: Open University Press.

Brookfield, H. C. 1964. Questions on the human frontiers of geography. *Economic Geography* **40**, 283–303.

Burgess, J. 1985. News from nowhere: the press, the riots and the myth of the inner city. In *Geography, the media and popular culture,* J. Burgess, & J.R. Gold, (eds.), 192–228. London: Croom Helm.

Burgess, J. & J. R. Gold, (eds.) 1985. *Geography, the media and popular culture.* London: Croom Helm.

Burgess, J. & P. Wood, 1988. Decoding Docklands: place advertising and the decision-making strategies of the small firm. In *Qualitative methods in human geography*, J. D. Eyles, & D. M. Smith, (eds.), 94–117. Cambridge: Polity Press.

Burgess, R. G. 1984. *In the field*. London: Allen & Unwin.

Burke, P. 1978. *Popular culture in early modern Europe*, New York: New York University Press.

Burrow, J. W. 1966. *Evolution and society: a study in Victorian social theory.* Cambridge: Cambridge University Press.

Carter, P. 1987. *The road to Botany Bay: an essay in spatial history*. London: Faber & Faber.

Cashmore, E. E. & B. Troyna, 1983. *An introduction to race relations*. London: Routledge & Kegan Paul.

Cassell, J. 1980. Ethical principles for conducting fieldwork. *American Anthropologist* **82**, 28–41.

Castells, M. 1976. Is there an urban sociology?. In *Urban sociology*, C. G. Pickvance (ed.), 33–59. New York: St. Martin's Press.

Castells, M. 1977. *The urban question*. London: Edward Arnold.

Castells, M. 1983. *The city and the grassroots*. London: Edward Arnold.

Cater, J. & T. Jones, 1987. Asian ethnicity, home-ownership and social reproduction. In *Race and racism*, P. Jackson, (ed.), 191–211. London: Allen & Unwin.

Chambers, I. 1986. *Popular culture: the metropolitan experience*. London: Methuen.

Chorley, R. J. & P. Haggett, (eds.) 1967. *Models in geography*. London: Methuen.

Clarke, J. 1984. 'There's no place like. . .': cultures of difference. In *Geography matters!*, D. Massey & J. Allen, (eds.), 54–67. Cambridge: Cambridge University Press.

Clarke, J. & C. Critcher, 1985. *The devil makes work: leisure in capitalist Britain.* London: Macmillan.

Clarke, J., C. Critcher, & R. Johnson, (eds.) 1979. *Working-class culture: studies in history and theory*. London: Hutchinson.

Clarke, J., S. Hall, T. Jefferson, & B. Roberts, 1976. Subcultures, cultures and class: a theoretical overview. In *Resistance through rituals*, S. Hall, & J. Henderson, (eds.), 9–74. London: Hutchinson/Centre for Contemporary Cultural Studies.

Clifford, J. 1986. Introduction: partial truths. In *Writing culture*, J. Clifford & G. E. Marcus, (eds.), 1–26. Berkeley, California: University of California Press.

Clifford, J. & G. E. Marcus, (eds.) 1986. *Writing culture*, Berkeley, California: University of California Press.

Cochrane, A. 1987. What a difference the place makes: the new structuralism of locality. *Antipode* **19**, 354–63.

Cockburn, C. 1983. *Brothers: male dominance and technological change*, London: Pluto Press.

Cockburn, C. 1986. The relations of technology: what implications for theories of sex and class?. In *Gender and stratification*, R. Crompton, & M. Mann, (eds.), 74–85. Cambridge: Polity Press.

Cohen, A. 1980. Drama and politics in the development of a London carnival. *Man* **15**, 65–87.

Cohen, A. 1982. A polyethnic London carnival as a contested cultural performance. *Ethnic and Racial Studies* **5**, 23–41.

Cohen, P. 1972. Sub–cultural conflict and working class community. *Working Papers in Cultural Studies* No. 2, Birmingham: Centre for Contemporary Cultural Studies, University of Birmingham.

Cohen, P. & C. Gardner, (eds.) 1982. *It ain't half racist, mum: fighting racism in the media*. London: Comedia.

Cohen, S. 1972. *Folk devils and moral panics*, London: MacGibbon & Kee.

Cohen, S. 1987. Symbols of trouble. Introduction to the new edition of *Folk devils and moral panics*, i-xxxiv. Oxford: Basil Blackwell.

Connell, R. W. 1985. Theorising gender. *Sociology* **19**, 260–72.

Connell, R. W. 1987. *Gender and power: society, the person and sexual politics*, Cambridge: Polity Press.

Cooke, P. 1986. The changing urban and regional system in the United Kingdom. *Regional Studies* **20**, 243–51.

Cooke, P. 1988. *Locality, structure and agency: a theoretical analysis*. Paper presented to the Annual Meeting of the Association of American Geographers, Phoenix, Arizona.

Cosgrove, D. E. 1983. Towards a radical cultural geography: problems of theory. *Antipode* **15**, 1–11.

Cosgrove, D. E. 1985a. Prospect, perspective and the evolution of the landscape idea. *Transactions, Institute of British Geographers* New series **10**, 45–62.

Cosgrove, D. E. 1985b. *Social formation and symbolic landscape*. London: Croom Helm.

Cosgrove, D. E. & S. J. Daniels, (eds.) 1988. *The iconography of landscape: essays on the symbolic representation, design and use of past environments*. Cambridge: Cambridge University Press.

Cosgrove, D. E. & P. Jackson, 1987. New directions in cultural geography. *Area* **19**, 95–101.

Crane, W. 1907. An artist's reminiscences. London: Methuen.

Cressey, P. G. 1932. *The taxi-dance hall*. Chicago: University of Chicago Press.

Cresswell, T. 1988. *Go-egraphy: mobility as resistance in modern American popular culture*. Paper presented to the Annual Meeting of the Association of American Geographers, Phoenix, Arizona.

Cross, M. 1982. The manufacture of marginality. In *Black youth in crisis*, E. Cashmore & B. Troyna (eds.), London: Allen & Unwin.

Cross, M. 1985. *Black workers, recession and economic restructuring in the West Midlands*. Paper presented to conference on 'Racial minorities, economic restructuring and urban decline'. Warwick: Centre for Research in Ethnic Relations, University of Warwick.

Dabydeen, D. 1985a. *Hogarth's blacks: images of blacks in eighteenth century English art*. Surrey: Dungaroo Press.

Dabydeen, D. (ed.) 1985b. *The black presence in English literature*. Manchester: Manchester University Press.

Damer, S. 1974. Wine Alley: the sociology of a dreadful enclosure. *Sociological Review* New series **22**, 221–48.

Daniels, S. J. 1981. Landscaping for a manufacturer: Humphry Repton's commission for Benjamin Gott at Armley in 1809–10. *Journal of Historical Geography* **7**, 379–96.

Daniels, S. J. 1982. Humphry Repton and the morality of landscape. In *Valued environments*, J. R. Gold & J. Burgess (eds.), 124–44. London: Allen & Unwin.

Daniels, S. J. 1985. Arguments for a humanistic geography. In *The future of geography*, R. J. Johnston, (ed.), 143–58. London: Methuen.

Darby, H. C. 1948. The regional geography of Thomas Hardy's Wessex. *Geographical Review* **38**, 426–43.

Darby, H. C. 1962. The problem of geographical description. *Transactions, Institute of British Geographers* **30**, 1–14.

Darnton, R. 1984. *The great cat massacre and other episodes in French cultural history*. New York: Basic Books.

Davenport, G. 1984. *The geography of the imagination*. London: Picador.

Dear, M. J. 1988. The postmodern challenge: reconstructing human geography. *Transactions. Institute of British Geographers* New Series, **13**, 262–74

de Blij, H. 1982. *Human geography: culture, society and space*, 2nd edn. Chichester: John Wiley.

Debord, G. 1983. *Society of the spectacle*. Detroit: Black and Red.

D'Emilio, J. 1983. *Sexual politics, sexual communities: the making of a homosexual minority in the United States*. Chicago: University of Chicago Press.

Dennis, R. J. 1984. *English industrial cities in the nineteenth century*. Cambridge: Cambridge University Press.

Donajgrodski, A. P. (ed.) 1977. *Social control in nineteenth century Britain*. London: Croom Helm.

Doughty, R. 1981. Environmental theology: trends and prospects in Christian thought. *Progress in Human Geography* **5**, 234–48.

Douglas, M. 1970. *Natural symbols: explorations in cosmology*. London: Barrie & Rockcliff.

Driver, F. 1985. Power, space, and the body: a critical assessment of Foucault's *Discipline and punish*. *Society and Space* **3**, 425–6.

Dummett, A. 1973. *A portrait of English racism*. Harmondsworth: Penguin.

Duncan, J. S. 1980. The superorganic in American cultural geography. *Annals, Association of American Geographers* **70**, 181–98.

Duncan, J. & N. Duncan, 1988. (Re)reading the landscape. *Society and Space* **6,** 117–26.

Duncan, J. S. & D. Ley, 1982. Structural Marxism and human geography. *Annals, Association of American Geographers* **72**, 30–59.

Eagleton, T. 1988. Resources for a journey of hope: the significance of Raymond Williams. *New Left Review* **168**, 3–11.

Elias, N. & Dunning, E. 1986. *Quest for excitement: sport and leisure in the civilizing process.* Oxford: Basil Blackwell.

Eliot, T. S. 1948. *Notes towards the definition of culture.* London: Faber & Faber.

Engels, F. 1884. *The origin of the family, private property and the state.* London: Lawrence & Wishart (1972 editions).

Engels, F. 1958. *The condition of the working class in England (1845).* New York: Macmillan.

Entrikin, J. N. 1984. Carl Sauer: philosopher in spite of himself. *Geographical Review* **74**, 387–408.

Fair, D. 1987. We the people, we the 25 million. *Gay Community News,* October 25–31.

Featherstone, M. 1988. In pursuit of the postmodern: an introduction. *Theory, Culture and Society* **5**, 195–215.

FitzGerald, F. 1987. *Cities on a hill: a journey through contemporary American cultures.* London: Picador.

Flett, H. 1979. Bureaucracy and ethnicity: notions of eligibility to public housing. In *Ethnicity at work* S. Wallman, (ed.), 135–52. London: Macmillan.

Foord, J. & N. Gregson, 1986. Patriarchy: towards a reconceptualisation. *Antipode* **18**, 186–211.

Foord, J., L. McDowell, & S. Bowlby, 1986. For love not money: gender relations in local areas. Discussion Paper No. 76. Newcastle: Centre for Urban and Regional Studies, Newcastle University.

Foucault, M. 1967. *Madness and civilisation.* London: Tavistock.

Foucault, M. 1972. *The archaeology of knowledge.* London: Tavistock.

Foucault, M. 1973. *The birth of the clinic.* London: Tavistock.

Foucault, M. 1976. *The history of sexuality, Vol. 1: an introduction.* Harmondsworth: Penguin.

Foucault, M. 1980. *Power/knowledge: selected interviews and other writings, 1972–1977*, C. Gordon (ed.). Brighton: Harvester Press.

Freeman, T. W. 1986. The unity of geography: an introduction. *Transactions, Institute of British Geographers* New series **11**, 441–2.

Fryer, P. 1984. *Staying power: the history of black people in Britain.* London: Pluto Press.

Geertz, C. 1973. *The interpretation of cultures.* New York: Basic Books.

Geertz, C. 1980. *Negara: the theatre state in nineteenth-century Bali.* Princeton, New Jersey: Princeton University Press.

Geertz, C. 1983. *Local knowledge.* New York: Basic Books.

Giddens, A. 1979. *Central problems in social theory.* London: Macmillan.

Giddens, A. 1985. *The constitution of society.* Cambridge: Polity Press.

Gilbert, A. 1988. The new regional geography in English and French-speaking countries. *Progress in Human Geography* **12**, 208–28.

Gilbert, E. W. 1972. British regional novelists and geography. In *British pioneers in geography*, E. W. Gilbert, (ed.), 116–27. Newton Abbott: David & Charles.

Gilbert, E. W. & R. W. Steel, 1945. Social geography and its place in colonial studies. *Geographical Journal* **106**, 118–31.

Gilroy, P. 1987. *'There ain't no black in the Union Jack': the cultural politics of race and nation*. London: Hutchinson.

GLC 1986. *A history of the black presence in London*. London: Greater London Council.

Gluckman, M. 1956. *Custom and conflict in Africa*. Oxford: Basil Blackwell.

Gramsci, A. 1971. *Selections from the prison notebooks*. London: Lawrence and Wishart.

Gramsci, A. 1973. *Letters from Prison*. New York: Harper & Row.

Gramsci, A. 1985. *Selections from the cultural writings*, London: Lawrence & Wishart.

Gregory, D. 1978. *Ideology, science and human geography*. London: Hutchinson.

Gregory, D. 1981. Human agency and human geography. *Transactions, Institute of British Geographers* New series **6**, 1–18.

Gregory, D. 1985. Suspended animation: the stasis of diffusion theory. In *Social relations and spatial structures*, D. Gregory, & J. Urry (eds.), 296–336. London: Macmillan.

Gregory, D. 1987. Postmodernism and the politics of social theory. *Society and Space* **13**, 130–54.

Gregory, D. 1989. Areal differentiation and post-modern human geography. In *New horizons in human geography*, D. Gregory, & R. Walford, (eds.) London: Macmillan, in press.

Gregory, D. & D. Ley, 1988. Culture's geographies. *Society and Space* **6**, 115–16.

Gregory, D. & J. Urry, (eds.) 1985. *Social relations and spatial structures*. London: Macmillan.

Hägerstrand, T. 1967. *Innovation diffusion as a spatial process*. Chicago: University of Chicago Press.

Haigh, M. J. 1982. The crisis in American geography. *Area* **14**, 185–9.

Hall, S. 1980a. Cultural studies and the Centre: some problematics and problems. In *Culture, media, language*, S. Hall, *et al.* (eds.), 15–47. London: Hutchinson/Centre for Contemporary Cultural Studies.

Hall, S. 1980b. Teaching race. *Multiracial Education* **9**, 3–13; reprinted in A. James & R. Jeffcoate, (eds.) 1981. *The school in the multicultural society*, 58–69. London: Harper & Row.

Hall, S. 1981. Notes on the deconstruction of 'the popular'. In *People's history and socialist theory*, R. Samuel, (ed.), 227–40. London: Routledge & Kegan Paul.

Hall, S. 1984. Reconstruction work. *Ten.8* **16**, 2–9.

Hall, S., C. Critcher, T. Jefferson, J. Clarke, & B. Roberts, 1978. *Policing the crisis: mugging, the state and law and order*. London: Macmillan.

Hall, S. & T. Jefferson, (eds.) 1976. *Resistance through rituals: youth subcultures in post-war Britain*. London: Hutchinson/Centre for Contemporary Cultural Studies.

Hargreaves, J. 1986. *Sport, power and culture: a social and historical analysis of popular sport in Britain*. Cambridge: Polity Press.

Harris, D. R. 1981. Breaking ground: agricultural origins and archaeological explanations. *Bulletin of the Institute of Archaeology* **18**, 1–20. London: University of London.

Harris, M. 1980. *Cultural materialism: the struggle for a science of culture.* New York: Random House.

Harris, R. 1980. *The language-makers,* London: Duckworth.

Harris, R. 1984. Residential segregation and class formation in the capitalist city: a review and directions for research. *Progress in Human Geography* **8**, 26–49.

Harrison, R. T. & D. N. Livingstone, 1982. Understanding in geography: structuring the subjective. In *Geography and the urban environment,* D. T. Herbert & R. J. Johnston, (eds.), Vol. 5, 1–39. London: Wiley & Sons.

Hart, J. F. 1982. The highest form of the geographer's art. *Annals, Association of American Geographers* **72**, 1–29.

Harvey, D. 1973. *Social justice and the city.* London: Edward Arnold.

Harvey, D. 1984. On the history and present condition of geography: an historical geography manifesto. *Professional Geographer* **36**, 1–11.

Harvey, D. 1985a *Consciousness and the urban experience.* Oxford: Basil Blackwell.

Harvey, D. 1985b. *The urbanization of capital.* Oxford: Basil Blackwell.

Harvey, D. 1987a. Three myths in search of a reality in urban studies. *Society and Space* **5**, 367–76.

Harvey, D. 1987b. Flexible accumulation through urbanisation: reflections on 'post-modernism' in the American city. *Antipode* **19**, 260–86.

Haug, W. F. 1986. *Critique of commodity aesthetics: appearance, sexuality and advertising in capitalist society.* Cambridge: Polity Press.

Hebdige, D. 1976. The meaning of mod. In *Resistance through rituals,* S. Hall, & T. Jefferson, (eds.), 87–96. London: Hutchinson/Centre for Contemporary Cultural Studies.

Hebdige, D. 1979. *Subculture: the meaning of style.* London: Methuen.

Hebdige, D. 1983. Travelling light: one route into material culture. *Royal Anthropological Institute News* **59**, 11–13.

Hebdige, D. 1987. *Cut 'n' mix: culture, identity and Caribbean music.* London: Comedia.

Hebdige, D. 1988. *Hiding in the light: on images and things.* London: Routledge, Chapman & Hall.

Hechter, M. 1975. *Internal colonialism: the Celtic fringe in British national development, 1536–1966.* London: Routledge & Kegan Paul.

Heger, H. 1980. *The men with the pink triangle.* London: Gay Men's Press.

Held, D. 1980. *An introduction to critical theory.* London: Hutchinson.

Henderson, J. & V. Karn, 1987. *Race, class and state housing: inequality and allocation of public housing in Britain.* London: Gower.

Hewison, R. 1987. *The heritage industry: Britain in a climate of decline.* London: Methuen.

Hicks, D. 1981. Bias in geography textbooks. *Working Paper* No. 1. London: Centre for Multicultural Education, University of London.

Hilton, T. 1970. *The Pre-Raphaelites.* London: Thames & Hudson.

Hoggart, R. 1957. *The uses of literacy: aspects of working class life with special reference to publications and entertainments.* London: Chatto & Windus.

Hoskins, W. G. 1955. *The making of the English landscape.* London: Hodder & Stoughton.

Hudson, B. J. 1982. The geographical imagination of Arnold Bennett, *Transactions, Institute of British Geographers* New series **7**, 365–79.
Humphries, J. 1977. Class struggle and the persistence of the working-class family. *Cambridge Journal of Economics* **1**, 241–58.
Humphries, J. 1981. Protective legislation, the capitalist state, and working class men: the case of the 1842 Mines Regulation Act, *Feminist Review* **7**, 1–33.
Humphreys, L. 1970. *Tearoom trade: a study of homosexual encounters in public places*, London: Duckworth.
Humphreys, L. 1972. *Out of the closets: the sociology of homosexual liberation*. Englewood Cliffs, New Jersey: Prentice-Hall.

Jackson, J. B. 1957–8. The abstract world of the hot-rodder. *Landscape* **7**, 22–7.
Jackson, J. B. 1976. The domestication of the garage. *Landscape* **20**, 10–19.
Jackson, J. B. 1984. *Discovering the vernacular landscape*. New Haven, Connecticut: Yale University Press.
Jackson, P. 1980. A plea for cultural geography. *Area* **12**, 110–13.
Jackson, P. 1985. Neighbourhood change in New York: the loft conversion process. *Tijdschrift voor Economische en Sociale Geografie* **76,** 202–15.
Jackson, P. 1987. The idea of 'race' and the geography of racism. In *Race and racism*, P. Jackson, (ed.), 3–21. London: Allen & Unwin.
Jackson, P. 1988a. Street life: the politics of Carnival. *Society and Space* **6**, 213–27.
Jackson, P. 1988b. Beneath the headlines: racism and reaction in contemporary Britain. *Geography* **73**, 202–7.
Jackson, P. & S. J. Smith, 1981. Introduction. In *Social interaction and ethnic segregation*, P. Jackson & S. J. Smith, (eds.), 1–17. London: Academic Press.
Jackson, P. & S. J. Smith, 1984. *Exploring social geography*, London: Allen & Unwin.
Jameson, F. 1984. Post-modernism, or the cultural logic of late capitalism. *New Left Review* **146**, 53–92.
Janik, A. & S. Toulmin, 1973. *Wittgenstein's Vienna*. New York: Simon & Schuster.
Janowitz, M. 1975. Sociological theory and social control. *American Journal of Sociology* **81**, 82–108.
Jay, A. O. M. 1891. *Life in darkest London*. London: Webster & Cable.
Jencks, C. A. 1981. *The language of post-modern architecture*. New York: Rizzoli.
Jones, G. 1980. *Social Darwinism and English thought: the interaction between biological and social theory*. Brighton: Harvester Press.
Jordan, T. G. 1978. *Texas log buildings: a folk architecture*. Austin, Texas: University of Texas Press.
Jordan, T. G. 1982. *Texas graveyards: a cultural legacy*. Austin, Texas: University of Texas Press.
Jordan, T. G. & L. A. Rowntree, 1982. *The human mosaic: a thematic introduction to cultural geography*. New York: Harper & Row.

Kaplan, C. 1986. *Sea changes: culture and feminism*. London: Verso.
Katznelson, I. 1981. *City trenches: urban politics and the patterning of class in the*

United States. Chicago: University of Chicago Press.

Keating, P. J. (ed.) 1976. *Into unknown England, 1866–1913: selections from the social explorers*. Manchester: Manchester University Press.

Keith, M. 1987. 'Something happened': the problems of explaining the 1980 and 1981 riots in British cities. In *Race and racism* P. Jackson, (ed.), 275–303. London: Allen & Unwin.

Kern, S. 1983. *The culture of time and space, 1880–1918*, Cambridge, Mass.: Harvard University Press.

Kerouac, J. 1958. *On the road*. London: Andre Deutsch.

Kerouac, J. 1960. *Lonesome traveller*. New York: McGraw-Hill.

Kerridge, R. 1983. *Real wicked, guy*. Oxford: Basil Blackwell.

Kniffen, F. B. 1979. Why folk housing?. *Annals, Association of American Geographers* **69**, 59–63.

Knopp, L. 1987. Social theory, social movements and public policy: recent accomplishments of the gay and lesbian movements in Minneapolis, Minnesota. *International Journal of Urban and Regional Research* **11**, 243–61.

Kolodny, A. 1975. *The lay of the land: metaphor as experience and history in American life and letters*. Chapel Hill, North Carolina: University of North Carolina Press.

Kolodny, A. 1984. *The land before her: fantasy and experience of the American frontiers, 1630–1860*. Chapel Hill, North Carolina: University of North Carolina Press.

Kroeber, A. L. 1917. The superorganic. *American Anthropologist* **19**, 163–213.

Kroeber, A. L. 1944. *The configurations of culture growth*. Berkeley, California: University of California Press.

Kroeber, A. L. 1952. *The nature of culture*. Chicago: University of Chicago Press.

Kroeber, A. L. & C. Kluckhohn, 1952. Culture: a critical review of concepts and definitions. *Papers of the Peabody Museum of American Archaeology and Ethnology* **47**, Cambridge, Mass.: Harvard University.

Kureishi, H. 1986. Bradford. *Granta* **20**, 149–70.

LaCapra, D. 1983. *Rethinking intellectual history: texts, contexts, language*. Ithaca, New York: Cornell University Press.

Laing, S. 1986. *Representations of working class life, 1957–1964*. London: Macmillan.

Lakoff, G. & M. Johnson, 1980. *Metaphors we live by*. Chicago: University of Chicago Press.

Lancaster Regionalism Group 1985. *Localities, class and gender*. London: Pion.

Larrain, J. 1979. *The concept of ideology*. London: Hutchinson.

Lash, S. & J. Urry, 1987. *The end of organized capitalism*. Cambridge: Polity Press.

Laska, S. B. & D. M. Spain, (eds.) 1980. *Back to the city: issues in neighborhood renovation*. New York: Pergamon.

Laslett, P. 1965. *The world we have lost*. London: Methuen.

Lauria, M. & L. Knopp, 1985. Toward an analysis of the role of gay communities in the urban renaissance. *Urban Geography* **6**, 152–69.

Lawrence, E. 1982a. 'Just plain common sense': the roots of racism. In *The*

Empire strikes back, Centre for Contemporary Cultural Studies, 47–94. London: Hutchinson.

Lawrence, E. 1982b. 'In the abundance of water the fool is thirsty': sociology and black 'pathology'. In *The Empire strikes back*, Centre for Contemporary Cultural Studies, 95–142. London: Hutchinson.

Leacock, E. B. (ed.) 1971. *The culture of poverty: a critique*. New York: Simon & Schuster.

Lee, D. & B. Loyd, 1982. *Women and geography: bibliography*. Cincinnati: Society of the Socially and Ecologically Responsible Geographers, University of Cincinnati.

Leighly, J. (ed.) 1967. *Land and life: a selection from the writings of Carl Ortwin Sauer*. Berkeley California: University of California Press.

Leighly, J. 1976. Carl Sauer (obituary). *Annals, Association of American Geographers* **66**, 337–48.

Levine, D. N. 1985. *The flight from ambiguity: essays in social and cultural theory*. Chicago: University of Chicago Press.

Levine, G. J. 1986. On the geography of religion. *Transactions, Institute of British Geographers*, New Series **11**, 428–40.

Levine, M. P. 1979. Gay ghetto. In *Gay men: the sociology of male homosexuality*, 182–204. New York: Harper & Row.

Lévi-Strauss, C. 1963. *Structural anthropology*. New York: Basic Books.

Lévi-Strauss, C. 1966. *The savage mind*. London: Weidenfield & Nicolson.

Lévi-Strauss, C. 1969. *The elementary structures of kinship*. Boston: Beacon Press.

Lewis, O. 1959. *Five families*. New York: Basic Books.

Lewis, O. 1961. *The children of Sanchez*. New York: Random House.

Lewis, O. 1964 *Pedro Martinez*, New York: Random House.

Lewis, O. 1965. *La vida*. New York: Random House.

Lewis, O. 1966. The culture of poverty, *Scientific American* **215** (4), 19–25.

Lewis, P. 1985. Beyond description. *Annals, Association of American Geographers* **75**, 465–77.

Ley, D. 1974. *The black inner city as frontier outpost*, Monograph No. 7. Washington, DC: Association of American Geographers.

Ley, D. 1981a. Behavioral geography and the philosophies of meaning. In *Behavioural problems in geography revisited*, K. R. Cox, & R. G. Golledge, (eds.), 209–30. New York: Methuen.

Ley, D. 1981b. Cultural/humanistic geography. *Progress in Human Geography* **5**, 249–57.

Ley, D. 1983. *A social geography of the city*. New York: Harper & Row.

Ley, D. 1986. Alternative explanations for inner city gentrification: a Canadian assessment. *Annals, Association of American Geographers* **76**, 521–35.

Ley, D. 1987. Styles of the times: liberal and neo-conservative landscapes in inner Vancouver, 1968–1986. *Journal of Historical Geography* **13**, 40–56.

Ley, D. 1988a. Landscape as spectacle: world's fairs and the culture of heroic consumption. *Society and Space* **6**, 191–212.

Ley, D. 1988b. From urban structure to urban landscape. *Urban Geography*, **9**, 98–105.

Ley, D. & R. Cybriwsky, 1974. Urban graffiti as territorial markers. *Annals, Association of American Geographers* **64**, 491–505.

Ley, D. & M. S. Samuels, (eds.) 1978. *Humanistic geography*. London: Croom Helm.

Little, J. Peake, L. & Richardson, P. (eds.) 1988. *Women in cities*. London: Macmillan.

Loach, L. 1984. We'll be here right to the end . . and after: women in the miners' strike. In *Digging deeper: issues in the miners' strike*, H. Beynon (ed.), 169–79. London: Verso.

Lowenthal, D. & M. J. Bowden, (eds.) 1976. *Geographies of the mind*. Oxford: Oxford University Press.

Loyd, B. 1982. *The prospering of San Francisco's gay neighborhoods*. Paper presented at the Annual Meeting of the Association of American Geographers in San Antonio, Texas.

Lyotard, J.-F. 1984. *The postmodern condition*. Manchester: Manchester University Press.

McDowell, L. 1979. Women in British geography. *Area* **11**, 151–4.

McDowell, L. 1983. Towards the understanding of the gender division of urban space. *Society and Space* **1**, 59–72.

McDowell, L. 1986. Beyond patriarchy: a class-based explanation of women's subordination, *Antipode* **18**, 311–21.

McDowell, L. & D. Massey, 1984. A woman's place. . . . In *Geography matters!* D. Massey, & J. Allen, (eds.), 128–47. Cambridge: Cambridge University Press.

McRobbie, A. & J. Garber, 1976. Girls and subcultures: an exploration. In *Resistance through rituals*, S. Hall, & T. Jefferson, (eds.), 209–22. London: Hutchinson.

Mailer, N. 1968. *The armies of the night*. New York: New American Library.

Mair, M. 1986. *Black rhythm and British reserve: interpretations of black musicality in British racist ideology since 1750*. Unpublished PhD dissertation, London: University of London.

Malinowski, B. 1922. *Argonauts of the Western Pacific*. London: Routledge & Kegan Paul.

Marcus, G. E. & M. M. J. Fischer (eds.) 1986. *Anthropology as cultural critique*. Chicago: University of Chicago Press.

Mark-Lawson, J. & A. Witz, 1986. *From 'family labour' to 'family wage'*. Working Paper No. 18, Lancaster: Lancaster Regionalism Group.

Marx, K. 1859. *Critique of Political Economy*. New York: International Publishers (1970 edition).

Marx, K. & F. Engels, 1846. *The German ideology*. London: Lawrence & Wishart (1970 edition).

Mass Observation 1986. *The pub and the people*. London: The Cresset Library.

Massey, D. 1984. *Spatial divisions of labour*. London: Macmillan.

Massey, D. 1985. New directions in space. In *Social relations and spatial structures*, D. Gregory, & J. Urry (eds.), 9–19. London: Macmillan.

Massey, D. 1988. A new class of geography. *Marxism Today*, May, 12–17.

Massey, D. & J. Allen (eds.) 1984. *Geography matters!*. Cambridge: Cambridge University Press.

Massey, D. & H. Wainwright, 1984. Beyond the coalfields: the work of the

miners' support groups. In *Digging deeper: issues in the miners' strike*, H. Beynon, (ed.), 149–68. London: Verso.
Meinig, D. W. (ed.) 1979. *The interpretation of ordinary landscapes*. Oxford: Oxford University Press.
Meinig, D. W. 1983. Geography as an art. *Transactions, Institute of British Geographers* New series **8**, 314–28.
Mikesell, M. W. 1977. Cultural geography. *Progress in Human Geography* **1**, 460–4.
Mikesell, M. W. 1978. Tradition and innovation in cultural geography. *Annals, Association of American Geographers* **68**, 1–16.
Mikesell, M. W. 1986. Sauer and 'Sauerology': a student's perspective. In *Carl O Sauer: a tribute*, M. S. Kenzer (ed.), 144–50. Corvallis, Oregon: Oregon State University Press.
Miles, R. 1982. *Racism and migrant labour*. London: Routledge & Kegan Paul.
Miles, R. 1984. The riots of 1958: the ideological construction of 'race relations' as a political issue in Britain. *Immigrants and Minorities* **3**, 252–75.
Miles, R. & A. Dunlop, 1987. Racism in Britain: the Scottish dimension. In *Race and racism*, P. Jackson, (ed.), 119–41. London: Allen & Unwin.
Millett, K. 1977. *Sexual politics*. London: Virago Press.
Momsen, J. H. & J. G. Townsend, (eds.) 1987. *Geography and gender in the Third World*. London: Hutchinson.
Morgan, L. H. 1877. *Ancient society*. Cambridge, Massachussetts: Harvard University Press (1964 edition).
Mort, F. 1988. Perfect lifestyle, ideal home. *Marxism Today*, March, 49–50.
Moser, C. O. N. & L. Peake (eds.) 1987. *Women, human settlements, and housing*. London: Tavistock.
Moynihan, D. P. 1965. *The Negro family: the case for national action*. Washington DC: US Department of Labor.
Murphy, R. 1983. A rival to Hollywood? The British film industry in the thirties. *Screen* **24**, 4–5.
Murray, B. 1984. *The old firm: sectarianism, sport and society*. Edinburgh: John Donaldson Publishers.

Nairn, T. 1988. *The enchanted glass: Britain and its monarchy*. London: Radius.
Neale, R. S. 1968. Class and class consciousness in early nineteenth-century England: three classes or five?. *Victorian Studies* **12**, 4–32.
Newton, J. L., M. P. Ryan, & J. R. Walkowitz, (eds.) 1983. *Sex and class in women's history*. London: Routledge & Kegan Paul.
New York City Police Department 1985. Rasta crime: a confidential report. *Caribbean Review* **14**, 12–40.
New York Department of City Planning 1984. *Private reinvestment and neighborhood change*. New York: Department of City Planning.
Norton, W. 1984. The meaning of culture in cultural geography: an appraisal. *Journal of Geography* **83**, 145–8.

Olsson, G. 1980. *Birds in egg, eggs in bird*. London: Pion.
OPCS 1987. *Developing questions on ethnicity and related topics for the Census*. Occasional Paper 36. London: Office of Population Censuses and Surveys.

Osofsky, G. 1966. *Harlem: the making of a ghetto*. New York: Harper & Row.
Owusu, K. & J. Ross, 1988. *Behind the masquerade: the story of Notting Hill Carnival*. London: Arts Media Group.

Pahl, R. E. 1984. *Divisions of labour*. Oxford: Basil Blackwell.
Pahl, R. E. (ed.) 1988. *On work: historical, comparative and theoretical approaches*. Oxford: Basil Blackwell.
Park, R. E. 1952. *Human communities*. Glencoe, Illinois: Free Press.
Parkin, F. 1979. *Marxism and class theory: a bourgeois critique*. London: Tavistock.
Parmar, P. 1984. Hateful contraries: media images of Asian women. *Ten.8* **16**, 71–8.
Parsons, J. J. 1976. Carl Sauer (obituary). *Geographical Review* **66**, 83–9.
Parsons, J. J. 1979. The later Sauer years. *Annals, Association of American Geographers* **69**, 9–15.
Paterson, J. H. 1965. The novelist and his region: Scotland through the eyes of Sir Walter Scott. *Scottish Geographical Magazine* **81**, 146–52.
Peach, C. 1968. *West Indian migration to Britain: a social geography*. Oxford: Oxford University Press for the Institute of Race Relations.
Peach, C. 1985. Immigrants and the 1981 urban riots in Britain. In *Contemporary studies of migration*, P. E. White, & G. A. van der Knapp, (eds.), 143–54. Norwich: GeoBooks.
Pearson, G. 1983. *Hooligan: a history of respectable fears*. London: Macmillan.
Peet, R. 1986. The destruction of regional cultures. In *A world in crisis? geographical perspectives*, R. J. Johnston & P. J. Taylor (eds.), 150–72. Oxford: Basil Blackwell.
Pepper, D. 1984. *The roots of modern environmentalism*. London: Croom Helm.
Pocock, D. C. D. (ed.) 1981. *Humanistic geography and literature*. London: Croom Helm.
Powell, E. 1978. *A nation or no nation?*. London: Batsford.
Pred, A. 1984. Place as historically contingent process: structuration and time-geography of becoming places. *Annals, Association of American Geographers* **74**, 279–97.
Prince, H. C. 1988. Art and agrarian change, 1710–1815. In *The iconography of landscape*, D. E. Cosgrove, & S. J. Daniels, (eds.), 98–118. Cambridge: Cambridge University Press.
Pudup, M. B. 1988. Arguments within regional geography. *Progress in Human Geography* **12**, 369–90.
Pugin, W. 1836. *Contrasts*, 2nd edn. 1841. London: Charles Dolman, (Reprinted by Leicester University Press, Victorian Library Ed. 1969).

Rainwater, L. & W. L. Yancey, (eds.) 1967. *The Moynihan report and the politics of controversy*. Cambridge, Mass.: MIT Press.
Redclift, N. & E. Mingione, (eds.) 1985. *Beyond employment: household, gender and subsistence*. Oxford: Basil Blackwell.
Riis, J. 1890. *How the other half lives*. New York: Hill & Wang (1957 edn).
Robins, D. & P. Cohen, 1978. *Knuckle sandwich: growing up in the working–class city*. Harmondsworth: Penguin.
Rooney, J. F. Jr., W. Zelinsky, & D. R. Louder, (eds.) 1982. *This remarkable*

continent. College Station, Texas: Texas A & M University Press for the Society of the North American Cultural Survey.
Rosaldo, M. Z. & L. Lamphere, (eds.) 1984. *Women, culture and society*, Stanford, California: Stanford University Press.
Said, E. W. 1978. *Orientalism*. New York: Pantheon Books.
Sarup, M. 1986. *The politics of multi-racial education*. London: Routledge & Kegan Paul.
Sauer, C. O. 1925. The morphology of landscape. *University of California Publications in Geography* **2**, 19–54.
Sauer, C. O. 1929. Historical geography and the western frontier. In *The trans-Mississipian West*, J. E. Willard & C. B. Goodykoontz (eds.), 267–89. Boulder, Colorado: University of Colorado.
Sauer, C. O. 1941. Foreword to historical geography. *Annals, Association of American Geographers* **31**, 1–24.
Sauer, C. O. 1952. *Agricultural origins and dispersals*, New York: American Geographical Society (reprinted as *Seeds, spades, hearths, and herds*, 1969, Cambridge, Mass.: MIT Press).
Sauer, C. O. 1956a. The education of a geographer. *Annals, Association of American Geographers* **46**, 287–99.
Sauer, C. O. 1956b. The agency of man on the Earth. In *Man's role in changing the face of the earth*. W. L. Thomas Jr. (ed.), 49–69. Chicago: University of Chicago Press.
Sauer, C. O. 1966. *The early Spanish Main*. Berkeley & Los Angeles, California: University of California Press.
Saunders, P. 1981. *Social theory and the urban question*, London: Hutchinson.
Saussure, F. de 1916. *Cours de linguistique générale*. Paris: Payot.
Sayer, A. 1985. The difference that space makes. In *Social relations and spatial structures*, D. Gregory & J. Urry, (eds.), 49–66. London: Macmillan.
Sayer, A. 1988. *Post-Fordism in question*. Paper presented to the Annual Meeting of the Association of American Geographers, Phoenix, Arizona.
Sayer, A. 1989. The 'new' regional geography and problems of narrative. *Society and Space* (forthcoming).
Scarman, Lord 1981. *The Brixton disorders of 10–12 April 1981*, Cmnd. 8427. London: HMSO.
Schorske, C. E. 1980. *Fin-de-siècle Vienna: politics and culture*. New York: Alfred A. Knopf.
Shankman, P. 1984. The thick and the thin: on the interpretive theoretical program of Clifford Geertz. *Current Anthropology* **25**, 261–79.
Shumsky, N. L. & L. M. Springer 1981. San Francisco's zone of prostitution, 1880–1934. *Journal of Historical Geography* **7**, 71–89.
Silk, J. 1984. Beyond geography and literature. *Society and Space* **2**, 151–78.
Simpson, C. R. 1979. *SoHo: the artist in the city*. Chicago: University of Chicago Press.
Sivanandan, A. 1983. Challenging racism: strategies for the '80s. *Race and Class* **25**, 1–11.
Smith, M. P. 1980. *The city and social theory*. Oxford: Basil Blackwell.
Smith, N. 1979a. Toward a theory of gentrification: a back to the city

movement by capital not people. *Journal of the American Planning Association* **45**, 538–48.

Smith, N. 1979b. Gentrification and capital: theory, practice and ideology in Society Hill. *Antipode* **11**, 24–35.

Smith, N. 1982. Gentrification and uneven development. *Economic Geography* **58**, 139–55.

Smith, N. 1984. *Uneven development*. Oxford: Basil Blackwell.

Smith, N. 1986. Gentrification, the frontier and the restructuring of urban space. In *Gentrification of the city*, N. Smith & P. Williams, (eds.), 15–34. London: Allen & Unwin.

Smith, N. 1987a. Dangers of the empirical turn. *Antipode* **19**, 59–68.

Smith, N. 1987b. Of yuppies and housing: gentrification, social restructuring, and the urban dream. *Society and Space* **5**, 151–72.

Smith, N. & P. Williams, (eds.) 1986. *Gentrification of the city*. London: Allen & Unwin.

Smith, S. J. 1984a. Negotiating ethnicity in an uncertain environment. *Ethnic and Racial Studies* **7**, 360–73.

Smith, S. J. 1984b. Crime and the structure of social relations. *Transactions, Institute of British Geographers* New series **9**, 427–42.

Smith, S. J. 1986. *Crime, space and society*. Cambridge: Cambridge University Press.

Soja, E. 1987. The postmodernization of geography: a review. *Annals, Association of American Geographers* **77**, 289–94.

Soja, E. W. 1988. *Postmodern geographies: the reassertion of space in social thought*. London: Verso.

Spear, A. H. 1967. *Black Chicago: the making of a Negro ghetto*. Chicago: University of Chicago Press.

Spencer, J. E. & W. L. Thomas, 1973. *Cultural geography: an evolutionary introduction to our humanized earth*, 2nd edn. New York: John Wiley & Sons.

Speth, W. W. 1981. Berkeley geography, 1923–32. In *The origins of academic geography in the United States* B. W. Blouet (ed.), 221–44. Hamden, Connecticut: Archon Books.

Stallybrass, P. & A. White, 1986. *The politics and poetics of transgression*. London: Methuen.

Stansell, C. 1986. *City of women: sex and class in New York, 1789–1860*. New York: Alfred A. Knopf.

Stedman Jones, G. 1971. *Outcast London: a study in the relationship between classes in Victorian society*, Oxford: Clarendon Press.

Stedman Jones, G. 1974. Working-class culture and working-class politics in London, 1870–1900: notes on the remaking of a working class. *Journal of Social History* **7**, 460–508.

Stedman Jones, G. 1983. *Languages of class*. Cambridge: Cambridge University Press.

Steiner, G. 1975. *After Babel: aspects of language and translation*. Oxford: Oxford University Press.

Steward, J. H. 1973. *Alfred Kroeber*. New York: Columbia University Press.

Stoddart, D. R. 1986. *On geography and its history*. Oxford: Basil Blackwell.

Storch, R. D. 1975. The plague of blue locusts: police reform and popular resistance in northern England, 1840–57. *International Review of Social History* **20**, 61–90.

Storch, R. D. 1976. The policeman as domestic missionary: urban discipline and popular culture in northern England, 1850–1880. *Journal of Social History* **9**, 481–509.

Stratton, J. 1977. *Pioneers in the urban wilderness.* New York: Urizen Books.

Street, B. 1975. *The savage in literature.* London: Routledge & Kegan Paul.

Street, J. 1986. *Rebel rock: the politics of popular music.* Oxford: Basil Blackwell.

Sumka, H. J. 1980. Federal antidisplacement policy in a context of urban decline. In *Back to the city*, S. B. Laska, & D. M. Spain, (eds.), 269–87. New York: Pergamon.

Susser, I. 1982. *Norman Street: poverty and politics in an urban neighbourhood.* Oxford: Oxford University Press.

Suttles, G. D. 1984. The cumulative texture of local urban culture. *American Journal of Sociology* **90**, 283–304.

Swann, Lord 1985. *Education for all*, Cmnd. 9453. Report of the Committee of Inquiry into the education of children from ethnic minority groups, London: HMSO.

Symanski, R. 1981. *The immoral landscape: female prostitution in Western societies.* Toronto: Butterworths.

Taub, R. P., D. G. Taylor, & J. D. Dunham, 1984. *Paths of neighborhood change: race and crime in urban America.* Chicago: University of Chicago Press.

Taueber, K. & A. Taeuber, 1965. *Negroes in cities.* Chicago: Aldine.

Taylor, I. 1971. 'Football mad': a speculative sociology of football hooliganism. In *The sociology of sport*, E. Dunning (ed.), 352–77. London: Frank Cass.

Taylor, I. 1982. On the sports violence question: soccer hooliganism revisited. In *Sport, culture and ideology*. J. Hargreaves, (ed.), 152–96. London: Routledge & Kegan Paul.

Thomas, K. 1964. Work and leisure in pre-industrial society, *Past and Present* **29**, 50–66.

Thomas, K. 1971. *Religion and the decline of magic.* London: Weidenfeld & Nicolson.

Thompson, E. P. 1955. *William Morris: romantic to revolutionary*. London: Lawrence & Wishart.

Thompson, E. P. 1963. *The making of the English working class.* Harmondsworth: Penguin.

Thompson, E. P. 1967. Time, work-discipline, and industrial capitalism. *Past and Present* **38**, 56–97.

Thompson, E. P. 1971. The moral economy of the English crowd in the eighteenth century. *Past and Present* **50**, 76–136.

Thompson, E. P. 1974. Patrician society, plebian culture. *Journal of Social History* **7**, 382–405.

Thompson, E. P. 1978. *The poverty of theory and other essays*, London: Merlin.

Thompson, F. M. L. 1981. Social control in Victorian Britain. *Economic History Review* **34**, 189–208.

Thompson, J. B. 1984. *Studies in the theory of ideology*, Cambridge: Polity Press.

Thrift, N. J. 1983. Literature, the production of culture and the politics of place. *Antipode* **15**, 12–23.

Thrift, N. J. 1989. Images of social change. In *Restructuring Britain*, Unit D314. Milton Keynes: Open University Press.

Tivers, J. 1978. How the other half lives: the geographical study of women. *Area* **10**, 302–6.

Tobier, E. 1979. Gentrification: the Manhattan story. *New York Affairs* **5**, 13–25.

Toll, R. C. 1974. *Blacking up: the minstrel show in nineteenth-century America.* Oxford: Oxford University Press.

Trindell, R. T. 1969. Franz Boas and American Geography. *Professional Geographer* **21**, 328–32.

Trudgill, P. 1975. Linguistic geography and geographical linguistics. *Progress in Human Geography* **7**, 227–52.

Trudgill, P. 1982. *On dialect: social and geographical perspectives.* Oxford: Basil Blackwell.

Trudgill, P. 1983. *Sociolinguistics: an introduction to language and society.* (revised edition.). London: Penguin.

Tuan, Yi-Fu 1978. Literature and geography: implications for geographical research. In *Humanistic geography*, D. Ley, & M. S. Samuels, (eds.), 194–206. London: Croom Helm.

Turner, V. 1967. *The forest of symbols.* Ithaca, New York: Cornell University Press.

Tylor, E. B. 1871. *Primitive culture.* London: John Murray.

Urry, J. 1981. *The anatomy of capitalist societies.* London: Macmillan.

Urry, J. 1988. Cultural change and contemporary holiday-making. *Theory, Culture and Society* **5**, 35–55.

Valentine, C. A. 1968. *Culture and poverty: critique and counter-proposals.* Chicago: University of Chicago Press.

Vidal, G. 1983. Sex is politics. In *Pink triangle and yellow star and other essays 1976–1982*, 188–209. London: Granada.

Wagner, P. L. 1975. The themes of cultural geography rethought. *Yearbook, Association of Pacific Coast Geographers* **37**, 7–14.

Wagner. P. L. & M. W. Mikesell, (eds.) 1962. *Readings in cultural geography.* Chicago: University of Chicago Press.

Walby, S. 1986. *Patriarchy at work: patriarchal and capitalist relations in employment, 1800–1984.* Cambridge: Polity Press.

Walvin, J. 1975. *The people's game: a social history of British football.* London: Allen Lane.

Walvin, J. 1982. Black caricature: the roots of racialism. In *'Race' in Britain: continuity and change.* C. Husband, (ed.) 50–72. London: Hutchinson University Library.

Walvin, J. 1986. *Football and the decline of Britain.* London: Macmillan.

Ward, D. 1976. The Victorian slum: an enduring myth?. *Annals, Association of American Geographers* **66**, 323–36.

Ward, D. 1984. The Progressives and the urban question. *Transactions, Institute of British Geographers* New series **9**, 299–314.

Ward, J. P. 1981. *Raymond Williams*. Cardiff: University of Wales Press.

Weatherford, J. M. 1986. *Porn row*. New York: Arbor House.

Weedon, C., A. Tolson, & F. Mort, 1980. Introduction to language studies at the Centre. In *Culture, media, language*, S. Hall *et al.*, (eds.), 177–85. London: Hutchinson/Centre for Contemporary Cultural Studies.

Weeks, J. 1977. *Coming out: homosexual politics in Britain from the nineteenth century to the present*. London: Quartet Books.

Weeks, J. 1981. *Sex, politics and society: the regulation of sexuality since 1800*. London: Longman.

Weeks, J. 1985. *Sexuality and its discontents: meanings, myths and modern sexualities*. London: Routledge & Kegan Paul.

Weightman, B. A. 1981. Towards a geography of the gay community. *Journal of Cultural Geography* **1**, 106–12.

White, E. 1986. *States of desire: travels in gay America*. London: Picador.

Widgery, D. 1986. *Beating time: riot 'n' race 'n' rock 'n' roll*. London: Chatto & Windus.

Wiener, M. J. 1981. *English culture and the decline of the industrial spirit, 1850–1980*. Cambridge: Cambridge University Press.

Williams, C. H. (ed). 1988. *Language in geographical context*. Clevedon, Philadelphia: Multilingual Matters Ltd.

Williams, M. 1983. 'The apple of my eye': Carl Sauer and historical geography. *Journal of Historical Geography* **9**, 1–28.

Williams, R. 1958. *Culture and society, 1780–1950*. London: Chatto & Windus.

Williams, R. 1960. *Border country*. London: Chatto & Windus.

Williams, R. 1961. *The long revolution*. London: Pelican.

Williams, R. 1962. *Communications*. Harmondsworth: Penguin.

Williams, R. 1973. *The country and the city*. London: Chatto & Windus.

Williams, R. 1974. *Television: technology and cultural form*. London: Fontana.

Williams, R. 1976. *Keywords: a vocabulary of culture and society*. London: Fontana.

Williams, R. 1977. *Marxism and literature*. Oxford: Oxford University Press.

Williams, R. 1979. *Politics and letters: interviews with New Left Review*. London: New Left Books.

Williams, R. 1980. *Problems in materialism and culture*. London: Verso.

Williams, R. 1981. *Culture*. London: Fontana.

Williams, S. W. 1983. The concept of culture and human geography: a reassessment. *Occasional Paper No. 5*. Keele: University of Keele, Department of Geography.

Willis, P. E. 1977. *Learning to labour: how working class kids get working class jobs*. London: Gower.

Willis, P. E. 1978. *Profane culture*. London: Routledge & Kegan Paul.

Wilson, W. J. 1987. *The truly disadvantaged: the inner city, the underclass, and public policy*. Chicago: University of Chicago Press.

Wirth, L. 1938. Urbanism as a way of life. *American Journal of Sociology* **44**, 1–24.

Women & Geography Study Group 1984. *Geography and gender: an introduction to feminist geography*. London: Hutchinson.

Woolf, P. 1988. Symbol of the Second Empire: cultural politics and the Paris Opera House. In *The iconography of landscape*, D. E. Cosgrove, & S. J.

Daniels, (eds.), 214–35. Cambridge: Cambridge University Press.
Wright, P. 1985. *On living in an old country*. London: Verso.

Yeo, E. & S. Yeo, (eds.) 1981. *Popular culture and class conflict, 1590–1914*. Brighton: Harvester.
Young, M. & P. Willmott, 1962. *Family and kinship in East London*. Harmondsworth: Penguin.

Zelinsky, W. 1973a. *The cultural geography of the United States*, Englewood Cliffs, New Jersey: Prentice Hall.
Zelinsky, W. 1973b. Women in geography: a brief factual account. *Professional Geographer* **25**, 151–65.
Zelinsky, W., J. Monk, & S. Hanson, 1982. Women and geography: a review and a prospectus. *Progress in Human Geography* **6**, 317–66.
Zelinsky, W. & C. H. Williams, 1988. The mapping of language in North America and the British Isles. *Progress in Human Geography* **12**, 337–68.
Zukin, S. 1986. Gentrification: culture and capital in the urban core. *Annual Review of Sociology* **13**, 129–47.
Zukin, S. 1988a. The postmodern debate over urban form. *Theory, Culture & Society* **5**, 431–46.
Zukin, S. 1988b. *Loft living: culture and capital in urban change*. London: Radius.

Index

[an 'n' indicates citation in the notes]

Printed in the United States
by Baker & Taylor Publisher Services